Science

&

Indigenous Traditional Knowledge

(Approaches of *Galo/Adi* tribes)

Kaushik Bhagawati
Kshitiz Kumar Shukla
Amit Sen
Rupankar Bhagawati

(kaushik.iasri@gmail.com)

2017

Published by

Authors
With due permission from:
ICAR Research Complex for NEH Region
Arunachal Pradesh Centre, Basar-791101
(kaushik.iasri@gmail.com)

{Dedicated to our beloved Miss. Kamakshi Bhagawati

10 months old daughter of K. Bhagawati}

Preface

Indigenous peoples and their traditional knowledge play critical roles in the collective global responses to the challenges posed changes in social structure and climate across the globe. The IPCC's Fourth Assessment Report (AR4) noted that 'indigenous knowledge is an invaluable basis for developing adaptation and natural resource management strategies in response to environmental and other forms of change'. Also, UNFCCC recognizes the conservation of traditional knowledge as co-benefits of ecosystem- based approaches to adaptation. They are critical to building appropriate local to global responses.

In this manuscript we present some significant and scientifically validated indigenous traditional knowledge and decision making approaches of indigenous *Galo/Adi* tribes of North Eastern Himalayan region of India related to farming and allied activities. *Galo/Adi* tribes are indigenous community of eastern part of Arunachal Pradesh. The indigenous traditional knowledge and approaches presented in this book are extracted from some research articles published by the authors in some national/international journals.

Authors

Acknowledgement

First of all we all bow our head in front of All Mighty God and our parents for giving us all the blessings and strength for the endeavour.

We specially thank the farmers of West Siang District of Arunachal Pradesh who indeed have deep understanding of environment and climate and for their help in the entire endeavour. We bow our head towards their rituals and customs which are scientific and may be beyond modern science. Thank you all.

We are thankful to all the staff of ICAR A.P. Centre, Basar for their support and help during the work.

Special thanks to Mrs. Anjumoni Mudoi Bhagawati, Mrs. Pratiksha Kshitiz Shukla and Mrs. Shama Sen for all their support and encouragement.

Thank you all

Authors

Contents

1

Introduction

Indigenous Traditional knowledge in this book mainly concern with the study of sources and the methods a society uses to obtain its food and other necessities keeping harmony with their immediate environment and ecology. The study is important particularly for evaluating human-ecology relationship and for further understanding of the human ecological relations to their environment.

If understood properly, the knowledge system can provide deeper insight into the many different aspects of sustainable development and the interrelated role of local peoples and their cultures. They have great potentials to strengthen the modern developmental programmes and make them sustainable and acceptable. It is our conviction that greater efforts should be made to identify this local specific knowledge, understand the rationale behind them, strengthen the adaptive capacity of local people by developing policies and projects with room for local knowledge base, and to develop methodologies that promote activities for improving livelihoods in a sustainable way.

All indigenous knowledge were collected from indigenous *Galo/Adi* farmers of West Siang District located in Eastern Himalayan region of Arunachal Pradesh.

The forests of the area are under traditional control of the indigenous tribal community and play significant important role in meeting the resource needs of the community. They have very close proximity with their environment and have intricate understanding of the natural phenomenons and processes. Their indigenous customs and myths, tradition and tales co-evolved with the nature and their environment. Forest resources played a vital role in their economy in the absence of commercial agriculture and industry.

The tribes of the region have an elaborate mosaic of cultures and strategies for subsistence that mainly revolve around and closely related to agriculture (particularly *jhum* or swidden cultivation) and animal husbandry. All the tribes and sub-tribes of the region practice agriculture both *jhum* and wet cultivation owing to two different types of terrain in Siang valley - steep slopes of the mountains and plain land of the foot hills. The *jhum* cultivation has been evolved in the long immemorial past in response to the difficult terrains and steep slopes of the region.

The world around us has been changing and would continue to change; these pro-nature systems increase our adaptation and resilience to these changes. Adaptation and resilience as a whole are gradual processes representing millions of ups and downs for a particular community or ecosystem. The approach of *Traditional Communities* reflects that their culture coevolved with environment to create a sustainable food procurement system.

The reasoning developed through continuous observations of certain phenomenon is the knowledge about the subject. The knowledge when transmitted to generations becomes "belief"

4

with experience. When we practice and live with this acquired knowledge, it becomes our culture which is our common beliefs and attitudes about certain event/thing/phenomenon sharing same location or environmental conditions. Thus observation is seed of knowledge and repeated observation refines it. Here we can say that it is observation and reasoning about certain subject or object followed by quest to find logic behind that reasoning through some repeatable steps is "Research".

With above factors as guiding principles, the main objective of this book is to document some of the basic traditional knowledge of the tribes that we, up to certain extent, able to correlate with the modern scientific knowledge.

Despite of wide acknowledgement of traditional knowledge systems, their incorporation in the programmes and policies related to modern scientific approaches and strategies are very limited. How to apply this indispensable knowledge in participatory research to these new realities of social, demographic and climate change needs to be explored. The decision-makers must base their policies and actions on best available knowledge.

2

Jhum Cultivation: Harmony with Nature and Climate

Summary

The knowledge behind the culture and beliefs of indigenous community needs to be harnessed and should be used to complement the modern technologies and policies for better and sustainable use of biological resources and increase resilience of the sector associated. The main objective of the current research was to study Jhum (Traditional Shifting Cultivation System) and the cycles and culture associated with it. The study was done in northeast

Himalayan region of India and phenomenological approach was used. The research reveals that Jhum is the component of traditional agro-ecosystem encompassing diverse set of knowledge and practices of indigenous and local communities embodying traditional life-styles relevant for the conservation and sustainable use of natural resources for their livelihood. The cycle associated with the system reflects the synergy of practices with the natural phenomenon and indicators. Contrary to common modern belief, Jhum is carbon sink, maintain soil health, preserve biological diversity and sustain local climate. Forest clearing during Jhum is not deforestation but forest modification allowing forest regrowth during sufficiently long fallow. Fundamentally, Jhum as a system is an integrated approach to establish agro-ecosystem in the difficult terrains of tropical hill regions that involve forest, soil, biodiversity and livestock management through their culture, tradition and rituals that coevolved with associated ecosystem. Instead of being threat to climate or environment, the system can provide deeper insight into the many different aspects of sustainable and climate resilient development; and the interrelated role of local peoples and their cultures.

INTRODUCTION

Science and scientific studies needs and demands diversity. No rule is the only rule of universe, no two events or things are exactly similar in the universe. There are no ideal or general conditions in nature; but full of diversity in all levels and locations. Something is there or something exists if we observe (sense) it and define it, and science begins if we try to understand the phenomenon behind it through continuous observation. The reasoning developed through these continuous observations is the knowledge about the subject. The knowledge when transmitted to generations becomes "belief" with experience. When we practice and live with this acquired knowledge, it becomes our culture which is our common beliefs and attitudes about certain event/thing/phenomenon sharing same location or environmental conditions [1]. Thus observation is seed of knowledge and repeated observation refines it. Here we can say that it is observation and reasoning about certain subject or object followed by quest to find logic behind that reasoning through some repeatable steps is

"Research". If science is knowledge about structure and behavior of the natural and physical world or study of universe and its phenomenon through observation and experiment [2], there should not be some specific rules or standards in science. To study such diverse world we need diverse context based approaches or methods, without following certain similar procedures. We cannot and should not define science in a box or bound it in some definite well define procedures and validate it by some statistical methods. The approaches need to be task or entity specific. What is visible or obvious may not be the fact all the time (when viewed without being the part of the system), in fact is not the fact in most of the time. In the pursuit of science or knowledge, the approach should be inclusive of all possible facet of the entity under investigation, and knowledge must also include know-how. Though experiments and statistics are required to support any findings but they should not be the determining factor. There are many facts and figures, may not be measurable or countable, but are the root of natural phenomenon and relationship. There should not be "no looking back" in science, we have to look back and if necessary we have to even go back. Go back to understand the event or entity in a new way or understand additional vital facts that we missed earlier, that created current crisis. Out of the innumerable examples, the environmental setback due to Industrial Revolution in Europe [3] and agricultural setback due to Green Revolution in India [4] were prominent.

Anthropogenic climate change and its impact is fact [5], though with spatially varied intensity, with evidences from increase in average air and ocean temperatures, melting of snow and ice and sea level rise. Now there is no need of any proof or declaration of climate change by some organization or institution; it can be felt and can be experienced. The very impact of the climate change in all the sectors, especially those that depend on ecosystem services are apparent. We now have to search the nodes where we missed the links and reassemble those links to define and shape our development in an environment friendly and sustainable way. One very important node is Indigenous Traditional Knowledge (ITK) which was neglected and tagged as primitive, but recently, it is getting much attention. The IPCC's

Fourth Assessment Report (AR4) [5] noted that 'indigenous knowledge is an invaluable basis for developing adaptation and natural resource management strategies in response to environmental and other forms of change' [6]. Also, UNFCCC (2011) [7] recognizes the conservation of traditional knowledge as co-benefits of ecosystem- based approaches to adaptation. The Institute of Advance Studies at the UN University recently identified more than 400 examples indicating the role of indigenous people in climate change monitoring, adaptation and mitigation [8]. The sustainability of the indigenous knowledge system lies on the fact that they were accumulated incrementally, tested by trial-and-error and transmitted to future generations orally or by shared practical experiences [9].

Here in this study, by highlighting ITK we are not trying to compare or condemn modern scientific systems but we are trying to find way out how science of ITK can be harnessed to complement the modern policies and strategies to make them more acceptable and increase participation. Especially for climate change mitigation and adaptation strategies, integration of indigenous knowledge can lead to the development of effective and robust strategies in a very cost-effective, participatory and sustainable way [10,11]. It is an attempt to find sense out of nonsense, science out of superstitions and knowledge out of culture. It is also an effort to demonstrate that this knowledge-practice-belief complex [12] is science blending both concrete and abstract. The study mainly focus on ITK associated with the *Jhum* (shifting cultivation) in its originality, being practiced mostly in tropical hilly areas of Southeast Asia, the Pacific, Latin America, the Caribbean, and Africa for millennia [13,14,15] and was estimated to be the means of subsistence for at least half a billion people [16]. Though *Jhum* has been continuously targeted as threat to climate [17,18,19,20,21], but in fact it is a promising prospect for climate change mitigation and adaptation. The validation process of modern science is based on scientific criteria that purportedly make distinction between useful from useless, objective from subjective, indigenous 'science' from indigenous 'beliefs'. Through this process, knowledge corresponding with the paradigm of Western science is extracted, and the rest is rejected [22]. This approach hurdles study of *Jhum* as a knowledge system that can provide resources for the future climate change mitigation and adaptation strategies. Before giving the rationale behind the current study, there needs to broaden the definition of certain terms under the given context.

Definition of '*Jhum*'

The word *Jhum* may be the most misunderstood term among the environmentalists, scientists and others concerned with natural resource management. The most commonly shifting cultivation is defined as any agricultural system in which the fields are cleared (usually by fire) and cultivated for shorter periods than they are fallowed [23]. *Jhum* is not synonymous with 'slash-burn agriculture' or it is not just a land-use system where forest is slashed and burned to make arable land for a particular period. In fact slush-burning is a land clearing method and is employed also for other agricultural and non-agricultural land clearing [24]. However, *Jhum* is a component of traditional agro-ecosystem encompassing practices derived from ages of observations to interact with the environment in most harmonious manner that took the form of traditions, customs and rituals that governs agriculture in a most cultural and sustainable way where land is cleared by means of controlled fire and employ natural fallow phase long enough to be dominated by woody vegetations. As pointed by Geist and Lambin (2001) [25], *Jhum* or traditional shifting cultivation (swidden-fallow farming) is very much different from shifting cultivation practiced by the migrant settlers (slash-and-burn agriculture by in-migrants). So it is very important to keep in mind that all shifting cultivation is not *Jhum*. And the most fundamental difference is that they are practiced by different group of people with wide variation in their culture and relationship with ecosystem.

Definition of 'Climate Change'

"Climate change in IPCC usage refers to a change in the state of the climate that can be identified (e.g. using statistical tests) by changes in the mean and/or the variability of its properties and that persists for an extended period, typically decades or longer" [5]. But for biological world, Climate Change is not just change in statistical distribution of weather patterns over a considerable period, but change in any factor in the immediate environment or global conditions that have impact on the stability of that particular ecosystem. It is not just change or variation in weather parameters like temperature, rainfall, wind etc but it must also include change in the immediate environment or ecosystem due to some new introduction of plants or/and animals, change in soil compositions, change in population & its distribution, change in perception & behaviour of the inhabitants etc. Climate in this sense is the sum total of factors that preserve the original identity of the location under consideration for a considerable period.

Jhum involves fire for clearing forest and consequently emits carbon into the atmosphere, thus considered to be detrimental to climate change [26,27] and is coming under increased pressure. It is largely viewed as an exploitative system involving poor management of natural resources and major factor behind deforestation, biodiversity loss and ecological instability. Its uniqueness from the rest of the cultivation practices in the plains by majority of the population account for it being widely misunderstood and misinterpreted. It is recognized as the practice of the primitive people and those who are outside the mainstream culture responsible for the wasteful use of resources [28]. However, the studies in the recent time agree that the effect of *Jhum* was greatly overestimated based on insufficient and erroneous information [29,30]. Also the role of *Jhum* is being recognized for ecologically sustainable and economically viable form of agriculture [30,31]. Though the practice was considered to be major factor for loss of biodiversity and imbalance in the ecosystem, but the fact remains that the majority of the world mega biodiversity area coincide with the area occupied by the indigenous people practicing similar system of agriculture from ages [32].

Current work never intended to nullify the findings and study of pervious research, but to draw attention towards some untouched and neglected aspects of *Jhum* cultivation that deserve through revision for sustainable management of ecosystem. It is to study the matrix i.e. the social, cultural and environment conditions under which *Jhum* developed. *Jhum* is not only an agricultural land use system for these indigenous peoples but is their way of life, with all their customs, rituals, festivals and heritage closely associated with it. The difficult topography, inhospitable terrain, incessant rains and harsh climatic conditions in the hilly regions led the people to adopt this age old practise. It is their cordial response to the difficulties in establishing an agro-ecosystem in the difficult tropical forest ecosystem that is characterised by generally poor acidic soils and a diverse flora & fauna leading to many potential competitor species for the food crops. The system was found to be economically and energetically efficient compared to other form of agriculture (terrace or valley cultivation) in heavy rainfall areas of the hill tracts [33]. The underlying science in selection of location, planting date, selection of crops & varieties, cultural management practices etc. warrant lot of research and study. The most noteworthy fact is though this slush-burning practice is being practiced from centauries in these hilly regions but still they have same evergreen diverse forest meeting all their resource need for livelihood right from house construction to medicines in a very sustainable way. The investigation is done to study the *Jhum* cultivation in general and peep into the science behind its sustainability, and the sustainability of ecosystem associated with it.

MATERIALS AND METHODS

Methodology

The phenomenological approach was used to study the various cultural system associated with *Jhum*. The approach was chosen principally because it evaluates and analyses natural behaviour and instinct, as the indigenous communities perceived it rather than imposing any sort of external value judgment [34]. The means and techniques through which the local people as "insider" come to know about some phenomenon can be investigated through this approach [35]. The approach also helps to differentiate *noumena* (things as they are) from *phenomena* (things as we perceive them). So first hand information was collected through interactions with farmers of various groups in their fields. The *Jhum* cultivation practices were closely studied right from land selection to storage. All the rituals and ceremonies related to the system were observed closely. Nearly 26 villages from 4 different states are covered during the investigation across the region and 166 numbers of farmers and other villagers were interviewed. The daily work calendar and cropping calendar of the different groups of farmers were collected and studied closely.

RESULTS

Jhum Calendar

The major stages of the traditional *Jhum* cycle in general are: site selection and clearing, burning, sowing, weeding, protection, harvesting and storage. Special and crucial decision concerning the location, scheduling, crops and the labour inputs needs to be taken in each stages of cultivation. This decision making process is very vital in the process and, though needs to take care of the agro-climatic and environmental conditions, and are also moulded by the social and cultural factors. The decision makers largely depend on the natural indicators for making vital decisions. Each tribe have some location-specific traditional calendar of events for *Jhum* with different local names. A distinction is made by the native tribes in the pattern of *Jhum* based on the locality where it is undertaken. Even within the same location, each site was found to have characterised by different time of sowing, harvesting etc. depending upon altitude of the site and vicinity to habitat. Further the fellow cycle is also very location-specific depending on nearby forest type and soil of the selected site.

Jhum cycle have a very interesting and nature-scientific relation with the natural indicators that worth further systematic study. Normally the *Jhum* cycle begins in the month of December-January with calls of some particular bird (*Chou pou*) or insects (*Goi*) or other location specific indicators. This cycle involves selection of new plots based on presence/absence of some selective vegetation. The selection criteria also depend on soil type of the area and nature of crop planned to grow. Some sites are considered to be sacred due to presence of some rare vegetation and likewise some areas nearby villages are considered to be cursed and cutting of trees in such area is strictly prohibited. Some tribes before felling of trees in common wastelands, they seek permission of deities. On one hand their mercy is sought for felling the tree and on the other they are thanked for rearing the tree for so long. During clearing of the forest they avoid cutting of some particular tree species so as not to invoke the spirits of the woods. A particular mention may be made of *Sengri & Sengne (Ficus sp.)* [36], which are considered to be the abode of sprits and to cut them or to use their wood as firewood is tabooed. It is followed by cutting and slashing of under growth, shrubs, twigs and trees. Cutting and Slashing are done by a selected group of skilled persons, and they use to cut trees of medium girth up to certain height keeping in mind the crop variety that they are planning to grow at that particular site and keep the height of slashed trees 6-8 inch lower than the expected height of the crop at its maturity. And also they were found to be uniformly distributed throughout the field. The next cycle is in the months of January-February with calls or coming of *Pipiar* birds or blooming of certain wild flower like *Bombax ceiba*, and the cycle involves burning of slash and clearing of charred remains. Very strict customs were followed while burning the slashed field in which the persons responsible are not supposed to take full meal and the event is celebrated overnight. The resulting ash was uniformly distributed throughout the field (Fig. 1). This stage also involves sowing of some early paddy and other location based crops. The months of February-March are the third cycle that begins with singing of *Pakyo tabo* bird and flowering of *Mekahi* (*Phoebe Cooperiana*). The main activities involve terracing of steep slopes and higher areas, along with contours with half burnt old logs, weeds, stems, etc. (Fig.2). Mainly sowing of maize is done during this period. In March-April, sowing of some vegetables like cucumber, cucurbits, chillies, ginger, beans, tapioca etc are done randomly mainly in the boundaries. Different tuber varieties are sown along the peripheries that act as live fence for protection against animals. *Mithuns (Bos frontalis)*, the most important domesticated herbivore animal is generally kept under temporary community confinement called *Lura* during the growing season [37]. Chirping of *Tuk pi pipi* and *pinching* birds starts the next cycle during the months of April-May that goes on till the time the frog starts croaking. Here the main activities involve weeding in pervious crops and sowing of paddy. The paddy is shown by dibbling techniques where farmers make a small hole with the help of a sharpen stick and drops two-three seed into the hole with expertise that they gain with years of experience and practice (Fig. 3). Sowing of paddy in some place also coincide with flowering of *Gynocardia odorata* known locally with different names. The sowing is mainly done by women folks and it involves minimum disturbance of soil. Almost all their festivals and rituals revolve around *Jhum* and they keep close monitoring over their field for weeding and other activities. The harvesting is done generally in the months of October-November in which the women plug the head of rice bunch and carry them in bamboo basket called *Egin*. The dried grains are stored in specially built rodent free granary called *Nehu*. Before using the new grains for food, they very religiously keep some portion of grains separately for future seed. Mixing seed of different variety is tabooed.

Cultural Practices

They keep close monitoring over the weather to schedule their activities of weeding and harvesting with the help of certain birds, insects and other natural indicators. An insect such as annual cicida is considered to be best for short range weather forecasting. With its unique singing pattern they can predict weather for next 2-4 days with near accuracy. Their sowing, harvesting and other vital activities are also significantly influenced by lunar cycle. Moon is considered the goddess of fertility and deity of flora & fauna. They schedule their activities, mainly sowing and harvesting, according to phases of moon. Days near the vicinity of Full Moon are considered ideal for sowing seeds to avoid insect attack and also they believe that the soil moisture is high during the days at the vicinity of Full Moon.

The indigenous *Jhumias* (*Jhum* cultivators) are very reluctant in adopting foreign crops and varieties. They have their own varieties of paddy,

33

maize and vegetables that they preserve from generations. Depending on the weather, site and soil, different varieties were sown. Preservation of seed for future crop is considered to be their spiritual responsibility and women folks are mainly responsible for the task. They generally have huge repository of germplasm of each crops.

A single site is generally sown for 3-5 years depending on the soil fertility of the site and nature of crops grown earlier. The minimum fallow duration is generally 15 years that too they decide after proper investigation of the vegetation type and stem diameter of the recovered plants.

DISCUSSION

The biodiversity conservation is always been in the forefront agenda of indigenous community practicing *Jhum*. However, *Jhum* was widely condemned as threat to biodiversity: both natural and agricultural [38]. The indigenous *Jhumias* maintain natural biological asset balance (the assets whose initial form was determined by ecosystem of the location) rather than giving more priority to some selected groups of biological assets. During the site selection for *Jhum* they strictly avoid those locations that are dominated by rare or/and medicinally important plants. They avoid felling certain big trees (regarding them as abode of sprit) as they knew its significance to immediate ecosystem and its sustainability. Such big trees also provide habitat to innumerable birds, animals and insects who are their guide in agricultural activities. Associating such practices with rituals and customs make them moral and spiritual responsibility of the society and individual. Site selection based on some indicator plants was also reported in Malaysia [39]. They worship the forest as protector; feeding and rearing human being [40]. Unlike modern strategies, the concrete and the spiritual co-exist side by side in the indigenous community, complementing and enriching rather than competing and contradicting [22]. The term 'Biological Diversity' encompasses virtually limitless subjects: cultural preservation, habitat preservation, species preservation etc [41]. *Jhum* is concerned not only with the preservation of flora-fauna, but those culture and traditions that evolved as a result of interdependence of the inhabitant with their immediate environment. It also induces understanding of nature and its phenomenon and their impact on society today and tomorrow. They have very cordial relationship with the seasonal birds, insects and animals; and are considered to be messenger of weather and fertility god, thus indiscriminate killing is strictly prohibited. The indigenous community socializes the natural phenomenon and social phenomenons are described in ecological terms [42]. The *Jhumias* consider themselves inside the system (ecosystem) which is by its nature a diverse system. Biodiversity is the major source of information and people gathers information by interacting with their immediate biological environment [43].

In the biologist's sense of the world, biological diversity is the natural stock of genetic material within an ecosystem [41]. The importance of genes lies on the fact that they determine the particular characteristics of a given organism and encode the information which determines the specific capabilities of that organism. Greater the varieties in the gene pool, greater are the variety of organisms, characteristics and traits. Beside forest biodiversity, agricultural bio-diversity is maintained by *Jhumias*. They preserve their original crop varieties in a very religious way. They also use to grow diverse crops or varieties as per land location and possible weather conditions. They have a huge repository of *germplasm* which they maintain as per their culture and tradition. Under the given environmental condition, a species is best to its own niche [44]. The region under investigation is the centre of origin of important crops like citrus, rice, etc. [45]. The crop wild relatives and landraces maintained by indigenous community have been considered to be essential to future viability of global food production irrespective of climate change [46]. It is the people and the practices associated with them, not the landscapes, conserve agricultural diversity. The hesitation of indigenous *Jhumias* in adopting foreign varieties is a great step towards biodiversity conservation because the uniform cultivated varieties that are now substituted for the resident diversity worldwide is posing major threat to biodiversity [41]. Studies found that the lake of adequate stock of the species from which it might regenerate itself has been the major reason behind any species being endangered [47]. Traditional management system like use of more varieties, species and landscape patches helps in conservation of biodiversity as found in several studies across the globe [48,49]. Ramakrishnan (1992) [33] describe mixed cultivation system especially in *Jhum*, in which some of the species help maintain ecosystem structure and function.

Grazing management through *Lura* is another innovative technique of biodiversity conservation of *Jhumias*, in which during the cropping and growing seasons all the *Mithuns* (*Bos frontalis*) of the community are temporarily confined in a sufficiently large selected site (that provide adequate food and water) which is changed every year. The confinement checks *Mithuns* from continuous, free and random grazing of forest vegetation during the growing season [37], besides protecting *Jhum* fields. During the growing and rainy season the confinement of *Mithuns* in *Lura* avoid disturbance of soil surface due to treading that check soil erosion and compaction; and allow free regeneration of grazed vegetations as well as seedling germination throughout the forest.

Jhum is mainly targeted as threat to climate change as it relies on fire [50,21], thus emitting carbon dioxide into atmosphere enhancing atmospheric concentration of greenhouse gases [27]. Fire and land & natural resource management are indispensible for indigenous people. Indigenous people promote diversity of habitat by regular burning of parts of ecosystem thereby increase stability and sustainability [51]. In fact fire has been used by the indigenous people as a tool to manage diverse ecosystem [52]. The *Jhumias* never go for reckless burning of slashed area. They know the science to control the fire and the ceremony keep them awake whole night so to avoid spread of fire. They avoid full meal during the time which may help them to avoid sleep. Also the ash increases the fertility of soil and help to lower the acidity of the soil which is the major constraints in the soils of hilly regions having high rainfall. Previous studies indicated that ash addition to the soil after burning mitigates the soil acidity and increases fertility [53,54,55].

Jhum is considered to be one of the major factors of deforestation and forest degradation [56,57]. But deforestation implies long-term (>10 years) or permanent loss of forest cover [58]. As per the definition of deforestation by FAO [59], the clearing of forest for *Jhum* is not deforestation but it is "forest modification". As indicated by van Noordwijk et al (2008) [60], clear-felling and replanting is normal forest management either naturally or artificially, as long as the re-growth is allowed to attain woody vegetation. On the contrary, by definition, if the slashed-burn land is converted for intensive agriculture it can be declared as deforestation.

The GHG emission (which is mainly carbon dioxide) during *Jhum* is not at all a luxury emission. Also if we consider whole budget of carbon dioxide (CO_2) emission of *Jhum*, the net emission is negligible [61,62]. Whatever is emitted during burning is nullified by land use, chemical-free managements, almost permanent land cover with alternate crops, non-flooded fields, forest regeneration in left out sites etc. The forest clearing for shifting cultivation releases less carbon than permanent forest clearing because fallow period allow forest re-growth [63]. It was found in the previous studies that if the fallow periods are long enough, shifting cultivation can be carbon neutral maintaining soil fertility [64,60]. Tillage can cause loss of significant amounts of carbon (lost as CO_2 bursts) immediately after tillage [65]. Reduced tillage or zero-tillage practices were found to decrease net emissions of carbon dioxide from soil [66,67] also retaining plant residue on the soil surface. The dibbling method of sowing by *Jhumias* is zero tillage emitting practically no carbon dioxide and retains soil organic matter (Fig. 3). Minimum disturbance of soil also lowers soil erosion and surface runoff. Soil erosion is also controlled by placing the half burn logs across the slopes of the field that lowers the speed of running rain water and retain the top soil (Fig. 2). Thus, *Jhum* also reduce CO_2 emission by avoiding soil erosion. Studies found that exposure of soil organic carbon to aeration during soil erosion increases CO_2 emissions. Soil erosion and degradation been never been an issue with *Jhum* cultivation though highlighted widely [68,69].

An over-emphasis on sequestering carbon in soil as a means of climate change mitigation may eclipse other issues that are at least as significant. One such issue is to identify ways to decrease emissions of non-CO_2 gases from agricultural practices, in view of the estimate by the IPCC (2007) [5] that 70% of the total GHG emissions from agriculture are associated with nitrogen fertilizer that released N_2O which has so and so times GHG potential compared to carbon dioxide. Chemical free cultivation and management practices results almost no emission of such gases during *Jhum*. Their system of cultivation is absolutely organic and natural, using only farm residues, animal wastes etc. Most of their insect and pest management strategies are based on performing field activities with natural phenomenon and indicators. They have very good understanding of insect pest and disease dynamics through ages of observations. They lower incidence of insect and pest by proper selection of date of sowing and other cultural management depending on lunar cycle. Sowing during the days in the vicinity of Full Moon found to lower insect attack and favour germination. Slashing medium girth plants to a certain height depending on the expected height of the crop to be shown is a unique insect control technique of *Jhumias*. During initial growth of the crop especially rice, the stumps acts as platform for birds to sit and feed on the insects in the leaf of crop, but when the plant grow to maturity it outgrow the stamps and avoid birds from feeding on its grains (Fig. 4). Burning of vegetations at *Jhum* sites besides adding carbon to the soil, also help neutralization of soil acidity. Soil acidity may be one of the main reasons of prevalence of diseases and pests in these areas, but burning not only control soil acidity but also help to get rid of spores of pests in the soil.

The *Jhumias* are very particular about fallow period and 15 years fallow was found to be sufficient for regeneration of vegetation and rejuvenation of soil. For the tropical forest the fallow period of minimum 10 years is generally sufficient for the recovery of the vegetation, but it depends on the nature of soil and vegetation [70]. The inspection of stem diameters of the recovered plants indicates the rejuvenation of the soil and the vegetation [39].

Thus the questions remains- why *Jhum* or other systems related to indigenous community were always targeted for any environmental degradation or crisis? Are the allegations having some logic or it is just shifting of blame? why our modern development policies and technologies in some way or other, now or then, here or there pose challenges to the natural environment or climate, while some age old indigenous practices are so sustainable and maintain diversity of ecosystem from antiquity? How we define development? Is development a uniform process? Where and what we are lacking fundamentally in our modern policies and strategies?

One of the most fundamental as well as visible implications of the modern human development has been the conversion of the naturally existing forms of assets to form that are preferred and valued most by the human societies [71]. Development become homogenous process over the past ten thousand years, where same developmental strategies were implemented across the globe where conversion was done by replacing the naturally existing slate of species with a selection from the same small menu of specialised species uniformly replacing the resident strategies and varieties. These lead to relative decline in diversity and resilience of the society. Domination and manipulation of nature by human activities become the definition of development. On the contrary the rituals and traditions of the indigenous people aims in conserving diversity in all aspects (biological, cultural, natural, social etc.) as it is *in situ*. The world around us has been changing and would continue to change; the pro-nature system increases our adaptation and resilience to these changes. Adaptation and resilience as a whole are gradual processes representing millions of ups and downs for a particular community or ecosystem [42]. The approach of *Jhumias* reflects that their culture coevolved with environment to create a sustainable food procurement system [72].

CONCLUSION

The indigenous inhabitants consider them as an indispensible part of the immediate environment and understand the role of every entity for its survival. They value the natural indicators like birds, trees, insects, animals and other natural phenomenon and follow their instinct to plan their activities. These help them to respond towards the irregularities of the weather and other natural events. The customs and traditions associated with *Jhum* cultivation phases safeguard their practices and systems leading to protection of climate. The deity or sprits are in fact some tools or mechanisms to implement their regulations over protecting certain landforms or some species that are vital for the ecosystem. All events and stages are marked by some symbols to associate some complicated events with the activity in a very logical or scientific manner. Biological diversity is not only a physical or environmental issue for *Jhumias*, but it also concern with diversity in thinking and perception. This very diversity helps native people to employ need-based strategies in case of adverse environmental and climatic conditions; and make them and their system resilient. Thus we can conclude that *Jhum* is prospect for sustainable development and climate change mitigation and adaptation. The structure of *Jhum* clearly depict that climate, ecosystem and species are one and the same representing a system, and change in any one of them leads to change in whole system. During the study it was observed that *Jhum* encompasses a complete body of knowledge, know-how and practices maintained and developed by peoples, generally agrarian in nature, who have extended histories of interaction with the natural environment. It provides the basis for local-level decision making about agriculture and adaptation to environmental or social change. If understood properly, instead of being threat to climate or environment, the system can provide deeper insight into the many different aspects of sustainable development and the interrelated role of local peoples and their cultures. It has great potential to strengthen the modern developmental programmes and make them sustainable and acceptable. It is our conviction that greater efforts should be made to identify the local specific knowledge, understand the rationale behind them, strengthen the adaptive capacity of local people by developing policies and projects with room for local knowledge base, and to develop methodologies that promote activities for improving livelihoods in a sustainable way.

REFERENCES

[1] Oxford Advanced Learners Dictionary. (2005). Oxford University Press, Great Claredon Street, Oxford, UK.

[2] http://www.oxforddictionaries.com/definition/english/science

[3] Mjøset, L. & Kasa, S. (1994). "Environmental problems and techno-economic paradigms, a contribution to the history of environmental problems". In R. Delorme & K. Dopfler (eds.): The Political Economy of Diversity, Edward Elgar Publishers, London, pp. 167–199.

[4] Singh, R.B. (2000). Environmental Consequences of Agricultural Development: a case study from the Green Revolution state of Haryana, India. Agriculture, Ecosystem and Environment, 82: 97-103.

[5] IPCC, Summary for Policymakers, Fourth Assessment Report (AR4). (2007). New York, Cambridge University Press

[6] Parry, M.L., Canziani, O.F., Palutikof, J.P., van der Linden, P.J. & Hanson C.E. (eds.) (2007). Contribution of Working Group II to the Fourth Assessment Report of the Intergovernmental Panel on Climate Change. Cambridge, UK and New York, Cambridge University Press.

[7] UNFCCC, Ecosystem-based approaches to adaptation: compilation of information. Note by the Secretariat.

Available at <
http://unfccc.int/resource/docs/2011/sbsta/eng/info
8.pdf> (2011)

[8]   Galloway McLean, K. (2010). Advance Guard: Climate
       change impacts, adaptation, mitigation and indigenous
       peoples. A compendium of case studies. UNU-IAS.

[9]   Ohmagari, K. & Berkes, F. (1997). Transmission of
       indigenous knowledge and bush skills among the
       Western James Bay Cree women of subarctic Canada.
       Human Ecology 25: 197-222

[10]   Hunn, E. (1993). What is traditional ecological
       knowledge? In: Williams N. and Baines G. (Eds.).
       Traditional ecological knowledge: Wisdom for
       sustainable development. Centre for Resource and
       Environmental Studies, ANU, Canberra, pp. 3-15.

[11]   Robinson, J. & Herbert, D. (2001). Integrating climate
       change and sustainable development. International
       Journal of Global Environmental Issues, 1(2), 130-148.

[12]   Berkes, F. (1999). Sacred ecology. Traditional
       ecological knowledge and resource management.
       Taylor and Francis, Philadelphia and London, UK.

[13]   Lawerence, D. & Schlesinger. W.H. (2001). Changes in
       soil phosphorus during 200 years of shifting
       cultivation in Indonesia, Ecology 82:2769-2780.

[14]   Eastmond, A. & Faust, B. (2006). Farmers, fires, and
       forest: a green alternative to shifting cultivation for
       conservation of the Maya forest? Landsc Urban Plan
       74:267-284.

[15] Thomaz, E.L. (2009). The influence of traditional steep land agricultural practices on runoff and soil loss, Agric. Ecosyst. Environ. 130:23-30.

[16] Craswell, E.T., Sajjapongse, A., Howlett, D.J.B. & Dowling, A.J. (1997). Agroforestry in the management of sloping lands in Asia and the Pacific. Agrofor Syst. 38:121–137

[17] FAO. (1957). Shifting Cultivation, Unasylva, 11:9-11.

[18] Dove, Michael R. (1983). Theories of swidden agriculture, and the political economy of ignorance. *Agroforestry Systems* 1: 85-99.

[19] IWGIA. (2007). Development-induced resettlement and social suffering in Laos. Indigenous Affairs 4/07. Copenhagen: IWGIA, p. 32, p. 26.

[20] Forsyth, T. & Andrew W. (2008). Forest Guardians, Forest Destroyers. The Politics of Environmental Knowledge in Northern Thailand. Chiang Mai, Thailand Silkworm Books

[21] Fox, J., Fujita, Y., Ngidang, D., Peluso, N., Potter, L., Sakuntaladewi, N., Sturgeon, J. & Thomas, D. (2009) Policies, Political-Economy and Swidden in Southeast Asia. Human Ecology, 37:305-322.

[22] Nakashima, D. & Roué, M. (2002). Social and economic dimensions of global environmental change, In: Encyclopedia of Global Environmental Change, P. Timmerman (Eds.), John Willey and Sons, pp- 314-324, Vol: 5.

[23] Conklin, H.C. (1957). Hanunoo Agriculture: a report on an integral system o f shifting cultivation in the Philippines. Rome: FAO (Forestry Development Paper no. 12

[24] Kerkhoff, E. & Sharma E. (2006). Debating Shifting Cultivation in the Eastern Himalayas:Farmers' Innovations as Lessons for Policy, Kathmandu: ICIMOD.

[25] Geist, H.J. & Lambin, E.F. (2001). What Drives Tropical Deforestation? A meta-analysis of proximate and underlying causes of deforestation based on subnational case study evidence. LUCC Report Series No. 4., 2001, Louvain-la-Neuve: CIACO.

[26] Brady N.C. (1996). Alternatives to slash-and-burn: a global imperative. Agriculture, Ecosystems & Environment, 58: p. 3-11.

[27] Rastogi, M., Singh, S. & Pathak, H. (2002). Emission of carbon dioxide from soil. Current Science, 82(5), 510-517.

[28] Warner, K. (1991). Shifting Cultivators, Local technical knowledge and natural resource management in humid tropics. Food and Agriculture Organization of the United Nations, Via delle Terme di Caracalla, 00100 Rome, Italy.

[29] Cairns, M. & Garrity, D.P. (1999). Improving shifting cultivation in Southeast Asia by building on indigenous fallow management strategies. Agrofor. Syst. 47:37-48.

[30] Brunn, T.B., de Neergaard, A., Lawrence, D. & Ziegler, A.D. (2009). Environmental consequences of the demise in swidden cultivation in Southeast Asia: carbon storage and soil quality. Humn. Ecol. 37:375-388.

[31] Ziegler, A.D., Bruun, T.B., Guardiola-Claramonte, M., Giambelluca, T.W., Lawrence, D., Lam, N.T. (2009). Environmental consequences of the demise in Swidden cultivation in montane mainland Southeast Asia: hydrology and geomorphology, Humn. Ecol. 37:361-373.

[32] Sobrevila, C. (2008). Role of Indigenous People in Biodiversity Conservation. The International Bank for Reconstruction and Development / THE WORLD BANK, 1818 H Street, N.W. Washington, D.C. 20433, U.S.A

[33] Ramakrishnan, P.S. (1992). Shifting Agriculture and Sustainable Development: an interdisciplinary study from north-eastern India. MAB Series, Volume 10, UNESCO, Paris.

[34] Mugati, T. & Maposa, R.S. (2012). Indigenous weather forecasting: A phenomenological study engaging the shone of Zimbabwe. The Journal of Pan African Studies, 4(9), 102-111.

[35] Cox, J.L. (1992). An Introduction to the Phenomenology of Religion, Gweru: Mambo Press.

[36] Gupta, V. (2004). Community forest management – A case study of East Kameng district, Arunachal Pradesh,

India, Consultancy report, CFMWG-NEI, NEHU, Shillong, India and Community Forestry International Inc., Santa Barbara, USA.

[37] Jini, D., Bhagawati, K., Singh, R., Bhagawati, R., Alone, R.A. & Ngachan, S.V. (2015). *Lura* – Indigenous Approach to Biodiversity Conservation by Temporary Community Confinement of *Mithuns* (*Bos frontalis*) during Growing Season. International Letters of Natural Sciences, 44:45-53. doi:10.18052/ www.scipress.com/ILNS.44.45.

[38] Terborgh, J. (1999). Requiem for nature. Island Press, Washington, D.C.

[39] Tanaka, S., Wasli, M.E., Seman, L., Jee, A., Kendawang, J.J. & Sakurai, K. (2007). Ecological study on site selection for shifting cultivation by the Iban of Sarawak, Malaysia. A case study in Mujong River area. Tropics, 16:357-372

[40] Nimachow, G. (2003). *Akas* and their forest: A study on traditional management of forest resources, Arunachal University Research Journal 6(2): 82.

[41] Swanson, T. (2013) Global Action for Biodiversity Conservation: An International Framework for Implementing the Convention on Biological Diversity, Earth Scan, London.

[42] Fairhead, J., and Leach, M. (1996). *Misreading the African landscape: society and ecology in a forestsavannah mosaic.* Cambridge University Press, Cambridge, UK.

[43] Iltis, H (1988). 'Serendipity in the Exploration of Biodiversity', in Wilson, E.O. (ed) Biodiversity, National Academy Press: Washington, DC.

[44] Eltringham, S.K. (1984).Wildlife Resources and Economic Development, John Wiley: New York

[45] Mao, A.A., Hynniewta, T.M. & Sanjappa, M. (2009). Plant wealth of Northeast India in reference to ethnobotany. Indian Journal of Traditional Knowledge, 8(1):96-103.

[46] Lane, A. & Jarvis, A. (2007). Changes in climate will modify the geography of crop suitability: Agricultural biodiversity can help with adaptation. Paper presented at ICRISAT/CGIAR 35th Anniversary Symposium, "Climate-Proofing Innovation for Poverty Reduction and Food Security", 22–24 November 2007, ICRISAT, Patancheru, India. Available at: http://www. icrisat.org/ Journal/ SpecialProject/ sp2.pdf.

[47] Clark, C.(1973) 'Profit Maximisation and the Extinction of Animal Species', Journal of Political Economy, 81(4):950-61

[48] Warren, D.M. (1995). Comments on articles by Arun Agrawal. Indigenous Knowledge and Development Monitor 4(1):13

[49] Sporrong, U. (1998). Dalecarlia in Central Sweden before 1800: a society of social and ecological resilience. Pages 67-94 in F Berkes and C. Folke, editors. Linking social and ecological systems: management practices and social mechanisms for

building resilience. Cambridge University Press, Cambridge, UK.

[50] Padoch, C., Coffey, K., Mertz, O., Leisz, S.J., Fox, J. & R.L. Wadley. (2007).The Demise of Swidden in Southeast Asia? Local Realities and Regional Ambiguities. Danish Journal of Geography 107(1): 29-41.

[51] Jackson, W.J. & Moore, P.F. (1998). The role of indigenous use of fire in forest management and conservation. International Seminar on Cultivating Forests: Alternative Forest Management Practices and Techniques for Community Forestry, Regional Community Forestry Training Center, Bangkok, Thailand.

[52] Lewis, H.T. (1989). Ecological and Technological Knowledge of Fire: Aborigines Versus Park Rangers in Northern Australia. American Anthropologist 91(4).

[53] Giardina, C.P., Sanford Jr, R.L,. Døckersmith, I.C. & Jaramillo, V.J. (2000).The effects of slash burning on ecosystem nutrients during the land preparation phase of shifting cultivation. Plant and Soil, 220: 247-260.

[54] Tanaka, S., Kendawang, J.J., Ishihara, J., Shibata, K., Kou, A., Jee, A., Ninomiya, I. & Sakurai, K. (2004). The effects of shifting cultivation on soil ecosystems in Sarawak, Malaysia. II. Changes in soil chemical properties and runoff water at Balai Ringin and Sabal Experimental sites. Soil Science and Plant Nutrition, 50: 689-699.

[55] Tanaka, S., Kendawang, J.J., Yoshida, N., Shibata, K., Jee, A., Tanaka, K., Ninomiya, I. & Sakurai, K. (2005). Effects of shifting cultivation on soil ecosystems in Sarawak, Malaysia. IV. Chemical properties of the soils and runoff water at Niah and Bakam experimental sites. Soil Science and Plant Nutrition, 51: 525-533.

[56] Kleinman, P. J. A., Pimentel D. & Bryant, R.B. (1995).The ecological sustainability of slash-and-burn agriculture. *Agriculture, Ecosystems and Environment*, 52(2-3): 235-249.

[57] Myers, N. (1992). The Primary Source: Tropical Forests and Our Future, W. W. Norton, New York.

[58] Giri, T. (2007). Strengthening Monitoring, Assessment and Reporting on Sustainable Forest Management in Asia (GCP/INT/988/JPN). Forest Department, Food and Agriculture Organization of the United Nation, Rome.

[59] Erni, C. (2009). Shifting the blame? Southeast Asia's indigenous peoples and shifting cultivation in the age of climate change. Paper presented at the seminar on "Adivasi/ST Communities in India: Development and Change", Delhi, August 27-29, 2009, p. 6

[60] Van Noordwijk M., Mulyoutami, E., Sakuntaladewi, N., & Agus, F. (2008). Swiddens in transition: shifted perceptions on shifting cultivators in Indonesia. Occasional Paper no.9. Bogor, Indonesia: World Agroforestry Centre.

[61] Fearnside, P.M. (2000). Global warming and tropical land-use change: Greenhouse gas emissions from biomass burning, decomposition and soils in forest conversion, shifting cultivation and secondary vegetation, Clim. Change, 46:115–158, doi:10.1023/A:1005569915357.

[62] Lehsten, V., Tansey, K., Balzter, H., Thonicke, K., Spessa, A., Weber, U., Smith, B. & Arneth. A. (2009). Estimating carbon emissions from African wildfires, Biogeosciences, 6: 349–360, doi:10.5194/bg-6-349-2009.

[63] Intergovernmental Panel on Climate Change (IPCC). (2006). Land Use, Land-Use Change and Forestry. IPCC, http://www.grida.no/ publications/other/ipcc_sr/ ?src=/climate/ipcc/land_use/008.htm#s6-1

[64] Ruthenberg, H. (1971). Farming Systems in the Tropics. Oxford: Clarendon Press.

[65] La Scala, N., Bolonhezi, D. & Pereira, G.T. (2006). Short-term soil CO_2 emission after conventional and reduced tillage of a no-till sugar cane area in southern Brazil. Soil Till. Res. 9:244-248.

[66] Noel, D.U. & Herby, B. (2000). Global climate change and the effect of conservation practices in US agriculture. Global Environmental Change 10:197-209.

[67] Wan, Y.F. & Lin, E.D. (2004). The influence of tillage on CH_4 and CO_2 emission flux in winter fallow cropland. Chinese Journal of Agrometeorology 3:8-10

[68] Gafur, A. (2001). Effects of shifting cultivation on soil properties, erosion, nutrient depletion and hydrological responses in small watershed of Chittagong Hill Tracts, unpublished PhD dissertation, Chemistry Department, The Royal Veterinary and Agricultural University, Copenhagen, Denmark.

[69] Gafur, A. Borggaard, O.K., Jensen, J.R. & Peterson, L. (2003) Run off and loses of soil and nutrients from watershed under shifting cultivation (*Jhum*) in the Chittagong Hill Tracts of Bangladesh. J. Hydrol., 279:293-309.

[70] Aryal, K.P. & Kerkhoff, E. (2008). The right to practice shifting cultivation as a traditional occupation in Nepal. A case study to apply ILO Conventions No. 111 (Employment and Occupation) and 169 (Indigenous and Tribal peoples) Kathmandu, International Labour Office.

[71] Solow, R. (1974). 'The Economics of Resources or the Resources of Economics', American Economic Review, 64:1-12.

[72] Gliessman, S.R. (1985). Economic and ecological factors in designing and managing sustainable agroecosystems. In Edens, T. C., Fridgen, C. and Battenfield, S. L. (eds.) Sustainable agriculture and integrated farming systems. East Lansing: Michigan State University Press, pp. 56 - 63.

*****

3

Lura – Indigenous approach to Biodiversity Conservation

Summary

The extraordinary knowledge of indigenous people about their immediate environment and natural resource base can be a great asset for conservation of biodiversity. The current study aims to investigate an indigenous method of grazing management through temporary confinement of Mithun (Bos frontalis) of whole village community in a well selected area in the forest during the cropping and growing season.

The whole system is called Lura and practiced by Galo tribes of Eastern Himalayan region of India. Every year Lura management committee is formed that selects a new site based on number of Mithuns, forage availability, time period and several other key criteria without affecting flora-fauna diversity and rare medicinal plants. The practice checks continuous, free, random and selective grazing by Mithuns. It prevent continuous disturbance of soil surface due to treading, during growing and rainy seasons that avoid soil erosion and compaction, and facilitate seedling germination and the invasion by plants. Change of site, provide resting period to the forages in the previous Lura site especially during growth stage that allow them to renew and regenerate appreciably within 1-2 months. It also saves resources and time for construction of fencing in each Jhum and other agricultural site of each farmer. The confinement offers easy monitoring, protection and regular health assessment of the livestock. Thus, it is a multifaceted indigenous practice that ensures grazing management, biodiversity conservation, protection of standing agricultural crops and animal health management.

INTRODUCTION

Eastern Himalayas (Indo-Myanmar) is a biodiversity hotspot. For effective and sustainable conservation of these natural stock of diversity, integrated, holistic and site specific approaches are needed. Majority of the area of this region coincide with the areas occupied by indigenous people. Indigenous people and biodiversity conservation are strongly and positively correlated [1]. Indigenous people are carriers of ancestral knowledge and wisdom about natural diversity and its sustainable management. Their effective participation in biodiversity conservation programs would result in more comprehensive and cost-effective conservation and management of biodiversity [1]. Out of various factor affecting ecology, animal overgrazing is a serious threat [2] leading to loss of biodiversity, irreversible loss of topsoil, increase of turbidity in surface water and increase in flooding frequency/intensity. For current study overgrazing is defined "as an excess of herbivory that leads to degradation of plant and soil resources" [3]. Also, here grazing is overgrazing when it conflict with the biodiversity conservation efforts [4]. Impacts of overgrazing to biodiversity and topsoil loss are of immense concern, since they are effectively irreversible. Species loss removes a resource that has a regeneration time of millions of years (it is primary output of four-and-a-half billion year evolutionary processes) [5], while significant topsoil loss has a regeneration time scale of tens of millennia [6]. The impact of overgrazing and related behavior on vegetation composition and species diversity were well documented [7, 8, 9, 10, 11]. Overgrazing can cause change or modification to plant morphology and physiology directly through defoliation and trampling, and indirectly through alteration in growth conditions [12]. In due course, overgrazing may cause directional change in the structure and composition of the plant communities due to alteration of the dynamics of plant population through impact of grazing intensity on species natality, density and mortality [13]. Herbivores have habit of selective grazing or browsing, usually preferring forages having high nutrient contents and low structural or chemical defenses [14]. Overgrazing decrease plant density of a particular species, which in turn favor growth of other plants that are less preferred by animals decreasing their food supply [15] and productivity.

Overgrazing is considered to be one of the major causes of soil degradation worldwide [16]. Prolong grazing on the same site changes soil physical, chemical and microbiological properties [17]. Inappropriate grazing practices leads to destruction of soil structure approved by high bulk density, high dry mechanical resistance and low structural stability [18]. Soil fertility largely depends on the presence of soil microorganisms, and their existence and activities in turn depends on soil water content and storage capacity, texture, size and rate of pores. The decrease in pore space and increase in bulk density, due to treading of animal, negatively affect soil microbes [19]. These especially become serious problem in hilly regions with high rainfall where the microorganism in soil is inherently low because of acidic nature of soil. Moreover, lack of ground cover due to overgrazing make the top soil susceptible to erosion [20]. Some studies indicate loss of soil carbon due to overgrazing [21].

Overgrazing does not depend on the number of animals, but it is the function of time that how long the animals are allowed to graze on a particular site. Overgrazing occur when the animals were kept in the same site for a considerable long period or are allowed to turn back to the same site before plants have recovered [22]. Degradation of the landscape may be a short term phenomenon and recovery is possible after grazing pressures have been greatly reduced. During peak growing period, periods for 1-2 months may be adequate for appreciable recovery. Overgrazing is a serious issue during the growing season as it has thrice more effect on the key forage species as compared to grazing during seasons when plants were senescent [23]. Thus grazing pressure depends on the season effect and overgrazing is result of grazing at inappropriate times relative to flora productivity cycle [24]. The defoliation of the forages during their growth period can reduce their vigor and their capacity to maintain growth. Seedling stage was identified to be the most critical stage [12] . Intensive grazing during growth period lowers the re-growth and renewing capacity of the plants [25], which is attributed of the fact that the level of non-structural carbohydrates reserve lowers due to defoliation [26]. Impact of grazing on biodiversity is a complicated and diverse issue that varies spatially and temporally [27] and management effort must be location and time specific.

The *Galo* tribes of West Siang District of Arunachal Pradesh (situated in north-eastern Himalayan region of India), have unique traditional methods for temporary confinement of *Mithun* (*Bos frontalis*) during cropping season called *Lura*. The region, owing to its diverse physical and climatic situations, harbor and sustain immense biodiversity (Indo-Myanmar zone). The livelihood of tribe is based on a natural resource that relies directly on biodiversity and ecosystem services, and they know that their livelihoods would be affected first and foremost by biodiversity loss. Conservation of biological diversity is in the forefront of their social agenda from antiquity through indigenous traditional methods specific to the location and its environment. These approaches provide resilience to their ecosystem, thus enabling them to adapt to changing condition. *Mithun* (*Bos frontalis*) is heavily built semi-domesticated bovine species originated in North Eastern Himalayan Region of India [28]. It is also found in Myanmar, Bhutan, Bangladesh as well as Tuman province of China.

Out of the total *Mithun* population (0.29 million) in India, 83.5 percent animals are found in Arunachal Pradesh [29].This animal is well adopted in steep forest at an elevation of 600 – 3000 meter above mean sea level (msl). *Mithun* have economic, social, cultural and religious significance to the Galo tribes. It is mainly a free-ranging forest dweller. The *Galo* tribes mainly practice *Jhum* cultivation (shifting cultivation) because of the geography and climate of the region. In *Jhum* cultivation each farmer of a community usually have large plot of land for cultivation. Management of *Mithun* is vital during cropping season to protect their crops, forest resources and also for their proper monitoring. The *Mithuns* of each village community were confined to a temporary selected enclosure in

75

forest during growing season. The site of *Lura* is selected based on certain key criteria. The site for confinement is changed every year. The objective of the current study was to investigate the various stages of *Lura* and to find its significance especially towards ecological sustainability.

METHODS USED

The study was conducted in *Lipu Namchi* village under Basar circle of West Siang district of Arunachal Pradesh. Arunachal Pradesh is situated in Eastern Himalayas as priority ecoregion and biologically rich 'hotspot' [30]. The state is custodian of 23.52 % of total flowering plants of India [31] and is also regarded as nature's repository of medicinal plants [32] where around 500 medicinal plants were identified during preliminary survey. The village is located N 28°

00.165' and E 094°40.832' at about 787-1150 meters (msl). The study site comes under alpine and temperate sub-alpine climate zone. The normal annual rainfall is around 2550 mm with mean rainy days of 144. The villages have 25 household with a population of around 158 consisting of *Galo* tribe. The village as a whole had 109 *Mithuns*. The study was carried from October 2012 to October 2014. The Extensive field survey was conducted along with the members of the *Lura* community. Detailed information was collected through focused group discussion with village elders and youths. The practices and associated rituals were closely studied in the *Lura* site.

LURA SYSTEM

Management Committee

Every year, before cropping season, the villagers constitute a *Lura* management committee comprising of village elders and youths, with a president, a vice-president and members. After sowing of paddy in the month of April, the committee conduct meeting at *Dera* (community hall) to make major decision regarding selection of site and other related activities associated with *Lura*. Each villager has to make necessary contribution for making confinement and activities related to monitoring and health care. Also they take active part in organization of rituals and ceremonies related to the practice.

Selection of Site

Site selection is a vital task of the *Lura*. Here in the region there is a unique land tenure system, the land belongs to community and not to the government, so selection of forest land is decision of the *Lura* committee. Since *Mithun* is free ranging, based on the number of animals the selected site must provide free room for the animal to gaze freely and to keep them physically fit. The selected site must contain adequate supply of forages and water for the period of 3-4 months. So the size of the site depends on number of *Mithuns*, available forages and period of confinement. Preference is given to site that required minimum fencing. *Mithuns* generally prefers deep and dense forest to avoid bright sunshine. The site must be easily reachable from the village and can be monitored properly. Beside these basic needs for the site, there are few very important aspects that are kept in priority while site selection. The selected site must not contain some rare and important species of plants or animals. Also it must not contain some plants of religious and medicinal importance. The marshy and shadowy areas are considered to be dwells of sprits, so are not suitable for the purpose. Several rituals and ceremonies are conducted in the area to take permission of the deity of forest and for well being of the animals. The selected forest area is either owned by certain family or families in the village or a community land. The selected area should be away from human dwelling to prevent inter-transmission of any infectious diseases during any epidemic as *Mithuns* are susceptible to

Tuberculosis, Para-tuberculosis, Brucellosis, Foot and Mouth disease (FMD), Infectious Bovine Rhinotracheitis (IBR), and Bovine Viral Diarrhea [33]. After the captivity period is over the reuse of the same site for next period is strictly avoided.

Construction of fencing

The construction usually starts from the month of May, at the start growing season of crops as well as forest. The area is fenced by the village community using locally available materials like bamboos, wooden post (10-15 cm diameter) and rope made of cane or bamboo. The height of the fence ranges from 1-1.5 m and provision of gates are made at eight or more different locations for entry and exit having a width 1-1.2 m. Temporary houses are constructed at various sites for monitoring and night watch. The entire process generally takes around 20-30 days for completion depending on the area of selected site. The bamboos and wooden posts used for the fencing are generally acquired during new moon day (or days near to it) to avoid or minimize insect attack. Various religious ceremonies and community feasts are organized during the process to seek blessing from their deity to protect their animal. After completion, the village priest called *Nyibu* chants prayers and grand feast is given for the entire village. *Mithun* owners are directed to bring their animals to the *Lura* as early as possible. Each *Mithun* kept inside the *Lura* have a unique identification ear mark which is distinct for a particular clan and family.

Monitoring

Beside protection of the agricultural crop, monitoring of *Mithuns* is one of the very important goals of the *Lura*. At least one person from each household, compulsorily, has to take part in the monitoring activity turn-wise and in rotation to form a group of 2-3 persons every day. The group periodically checks and monitors the condition of fences, health of *Mithun*, intrusion of any predators etc. They provide salt and medicines at certain interval to all the animals. During the period of captivity human movements except the selected members are restricted in *Lura*. In case of injury or disease outbreak they inform the committee to take necessary suggestion of village elders or Veterinary Department. In case of calving the owners is informed who provide salt to the calves and brings it along with the mother to their houses for proper health care and making identification marks by doing ear notching with an sharp knife and sterilization by applying wooden ash till wound is healed [34]. The committee regularly repairs the fences at an interval 20-30 days.

Natural feed resources in LURA

The *Mithuns* browse on wide varieties of natural fodders *viz.* tree leaves, herbs, shrubs, grasses and creepers available. The common feed resources identified by farmers were: 1) Tree fodders (*Ficus hirta, Ficus* spp., *Oreocnide integrifolia, Sarcochlarvys pulcherrima* and *Bischofia javanica*), 2) Shrubs and Herbs (*Musa* spp., *Saccharum spontaneum, Boehmeria* spp. and *Dendrocalamus hamiltonii*), 3) Creepers

(*Conocephalus sauvelence*, *Puereria* spp., *Entada poseatha* and *Micrenga micranta*) and 4) Grasses (*Setaria palmifolia* and *Carex cruciata*). They selectively browsed on delicate leaves and accessible twigs of the branches. Bamboo leaves and their shoots were selectively preferred. The *Mithuns* feed more for the *Musa* spp. during hot part of the day and delicate leaves were preferred over to mature or dry leaves.

The gates are opened after the harvesting of paddy field in the month of September-October.

COMMENTS

Lura may be regarded as farmers' innovation for conservation of bio-diversity, social governance system and agricultural & veterinary management system. The confinement checks *Mithuns* from continuous, free and random grazing of forest vegetation during the growing season. The seedling germination and plant growth is inhabited when the soil surface is continuously disturbed by animals during growing season [35]. Though in the area, *Mithun* density is adequate for forage yield of the forest adjacent to the villages, they are not allowed to graze freely especially during growing season. This might be due to the fact that domestic livestock are usually driven by habit like preference for particular vegetation and preference for previously grazed area. Various literatures have indicated that livestock normally select and continuously graze on most preferred and palatable plant species first, leading to the death of the plant due to complete defoliation [36]. As when animals are allowed to graze freely, overgrazing generally occurs in same pasture and on same forage species [37]. Moreover, observation on Hungarian Grey Cattle reveal that though pasture yield was sufficient for animal density, difference in utilization leads to overgrazing in some part [38]. The *Lura* site is changed every year to allow the diverse high-quality and *mithun* preferred grasses and other forages to re-grow and renew in the previous site during growing season and to thrive sustainably, particularly the taller growing forages (trees, shrubs and herbs) that usually die under continuous grazing. It is in accordance to suggestion of Pratt (2002) [22] to move livestock out of a pasture before re-growth begins to prevent overgrazing. The farmers prefer to maintain diverse plant communities rather than going for some specific forage type. This might be due to the fact that diverse plant communities are more resilient and resistant to disturbances, which is in accordance to findings of Marañón (1997) [39].

The confinement of Mithuns during growing season in *Lura* might avoid disturbance of soil surface, especially in the left out *Jhum* slopes and previous *Lura* site, thus check soil erosion and compaction and allow free regeneration of grazed vegetations as well as seedling germination throughout the forest. Similar conclusion was drawn from the study on deer grazing in New Zealand indicating that continuous grazing can cause compact of topsoil and destruction of soil structure [40. The confinement of the livestock during growing season that coincide with period of heavy rainfall also checks indiscriminate treading. It was found that scars left by sheep disturbed disproportionately more during growing season hindering regeneration of the bare soil [41]. Also, livestock grazing intensity and lack of vegetation cover affect the soil porosity [42] leading to high bulk density. The annual shifting of *Lura* site might allows recovery of the compaction of soil caused by treading in the previous *Lura*. Change of site, also might allow soil to renew and regenerate appreciably within 3-4 months. Lack of continuous grazing pressure allows soil to improve and compensate quickly [43]. During *Lura*, grazing systems cause livestock to graze more uniformly and completely. This enhance metabolism of nutrients to growing points per root biomass [44], thus facilitate fast recovery of pasture during rest period.

Beside bio-diversity conservation, the system also provides an effective method to saves the standing agricultural crops. It also saves resources and time for construction of fencing in each *Jhum* and other agricultural site of each farmer. It also preserves the rare and important medicinal plants species during growing season. Monitoring of the animal against seasonal and infectious diseases is another important aspect of *Lura*, along with protection from predators. The farmers can also divert their attention towards farming without bothering whereabouts and safety of their *Mithuns*. The customs and ceremonies associated with *Lura*, besides strengthening the unity of community, make the system rigid and mandatory. Overall *Lura* is an all inclusive traditional practice that has multifaceted benefits that maintain a harmony and ecological balance between human, animals and nature. Biodiversity conservation with the participation of indigenous people presents great opportunities as they have extraordinary knowledge of their immediate environment.

Limitation in the system includes the initial cost and labor involved in fencing of such vast area, and labor for constant monitoring.

REFERENCES

[1]   C. Sobrevila. Role of Indigenous People in Biodiversity Conservation. The International Bank for Reconstruction and Development / THE WORLD BANK, 1818 H Street, N.W. Washington, D.C. 20433, U.S.A, 2008.

[2]   S.G. Bazha. Contemporary processes of degradation of pastoral steppe ecosystems in Mongolia. Proceedings of International conference. Ecological consequences of Biosphere process in the ecotone zone of southern Siberia and Central Asia, Ulaanbaatar, Mongolia, (2010).

[3] M.B. Coughenour, F.J Singer. The concept of overgrazing and its application to Yellowstone's northern range, In: R.B. Keiter, M.S. Boyce (Eds.); The Greater Yellowstone ecosystem. Redefining America' Wilderness Heritage. Yale University Press, New Haven, CT, 1991, pp-29-30.

[4] A. Mysterud. The concept of overgrazing and its role in management of large harbivores. Wildlife Biology 12 (2006)129-141.

[5] T. Swanson. Global Action for Biodiversity Conservation: An International Framework for Implementing the Convention on Biological Diversity, Earth Scan, London, 2013.

[6] C. Hogan. Overgrazing. (2012) www.ecoearth. org/ view/article/155088

[7] R.L. Jefferis, D.R. Klein, G.R. Shaver. Vertebrate herbivores and northern plant communities: reciprocal influences and responses, Oikos, 71 (1994) 193-206

[8] N.T. Hobbs. Modification of ecosystems by ungulates. Journal of Wildlife Management, 60 (1996) 695-713

[9] D.J. Augustine, S.J. McNaughton. Ungulate effects on the functional species composition of plant communities: herbivore selectivity and plant tolerance, Journal of Wildlife Management, 62 (1998) 1165-1183

[10] J.P. Bakker. The impact of grazing on plant communities. In: M.F. WallisDe Vries, J.P. Bakker, S.E. Van Wieren (Eds), Grazing and conservation management. Kluwer, Dordrecht, The Netherlands, 1998, pp. 137-185

[11] G. Austrheim, O. Eriksson. Plant species diversity and grazing in the Scandinavian mountains-patterns and processed at different spatial scales, Ecography. 24 (2001) 683-695.

[12] L. Torrano, J. Valderfabano. 2005. Review. Impact of grazing on plant communities in forestry areas. Spanish Journal of Agricultural Research, 2(1) (2005) 93-105

[13] S. Archer, F.E. Smeins. Ecosystem-level processes. In: R.K. Heitschmidt, J.W. Stuth (Eds.), Grazing management. An ecological perspective Timber Press, Portland, Oregon, USA, 1991, pp. 109-139.

[14] T.A. Hanley. A nutritional view of understanding and complexity in the problem for diet selection by deer (Cervidae), Oikos, 79 (1997) 209-218.

[15] PAI (Plant-Animal Interactions). Colorado State University Cooperative Extension. http://www.coopext.colostate.edu/SEA/Tim/plant-animal.htm. (2004)

[16] L.R. Oldemann, R.T.A. Hakkeling, W.C. Sombroek. World Map of the Status of Human-induced Soil Degradation: An Explanatory Note,

93

2nd revised edn. International Soil Reference and Information Centre, Nairobi/United Nations Environment Programme, Wageningen (1991)

[17] M.A. Steffens, Kölbl, K.U. Totsche, I. Kögel-Knabner. Grazing effects on soil chemical and physical properties in a semiarid steppe of Inner Mongolia (P.R. China), Geoderma, 143 (2008) 63–72.

[18] M.B. Villamil M. B., N. M. Amiotti & N. Peinemann. Physical fertility loss of soils in the southern Caldenal region, (Argentina), by overgrazing. Ciencia-del-Suelo. 15 (1997) 2. 102-104.

[19] Kátai J. Relationships between the physical, chemical and microbiological characteristics on a grassland experiment. Proc. of the 17th General Meeting of the EGF, Debr, 1998, 77-81.

[20] E. Mapfumo, D.S. Chanasyk, V.S. Baron, M.A. Naeth. Grazing impacts on selected soil parameters under short-term forage sequences, J. of Range Manage, 53 (2000) 466-470.

[21] LingHao, L. Ch. ZuoZhong, Q. QiBing, L. XianHua, L. YongHong, L. H. Li, Z. Z. Chen, O. B. Wang & X. H. Liu. Changes in soil carbon storage due to over-grazing in Leymus chinensis steppe in the Xilin River Basin of Inner

94

Mongolia. Journal of Environmental Sciences 9(4) (1997) 486-490

[22] D. Pratt. Stop Overgrazing. Beef. Minneapolis. 38 (2002)12. 22.

[23] D.N. Hyder, R. E. Bement, E. E. Remmenga & D. F. Hervey. Ecological responses of native plants and guidelines for management of shortgrass range. United States Department of Agriculture-Agricultural Res. Service, Tech. Bulletin Number 1503, US Government Printing Office, Washington, D. C. 87 (1975)

[24] D. Price, D. "What is Overgrazing?" Beef (periodical), Primedia Business Magazines, May 1(1999)

[25] R.A. Gregory, P.M. Wargo. Timing of defoliation and its effect on bud development, starch reserves and sap sugar concentration in sugar maple. Canadian Journal of Forest Research, 16 (1986) 10-17.

[26] S.A. Grant, G.T. Barthram, W.I.C. Lamb, J.A. Milne. Effects of season and level on grazing on the utilization of heather by sheep. 1. Responses of the sward. Journal of Brazilian Grassland Society, 33 (1978) 289-300.

[27] G.R. Miller, C. Geddes, D.K. Mordon. Response of the alpine gentran Gentiana nivalis L. to protection from grazing by sheep, Biological Conservation 87 (1999) 311-318

95

[28] M. Mondal, A. Dhali, C. Rajkhowa, B.S. Prakash. Secretion Patterns of Growth Hormone in Growing Captive Mithuns (Bos frontalis). Zoological Science. 21 (2004) 1125-1129.

[29] Livestock Census (19th). Ministry of Agriculture, Government of India. Department of Animal Husbandry, Dairying and Fisheries, Krishi Bhawan, New Delhi, 2012.

[30] WWF Annual Report. WWF International Avenue du Mont-Blanc 1196 Gland Switzerland, 2005, Pp: 24

[31] S.N. Hegde. Arunachal Pradesh State Biodiveristy Stretagy & Action Plan - Final Report. State Forest Research Institute, Itanagar, 2002.

[32] K. Haridasan, G.P. Shukla, B. S. Benewal. Medicinal Plants of Arunachal Pradesh. SFRI Information Bulletin, No.5. State Forest Research Institute, Itanagar, 1989.

[33] S. Rajkhowa, C. Rajkhowa, K.M. Bajrbauah. Diseases of mithun (Bos Frontalis) a revised Veterinary Bulletin 73 (2003) IR-6R.

[34] T. Heli, S. Saikia. Mithun: The pride of Arunachal Pradesh. Indian Farming, 57(5) (1996) 33-35

[35] J.H. Tallis, D.W. Yalden. Peak District Moorland Restoration Project: Phase 2 Report. RP-

vegetation Trials. Peak Park Joint Planning
Board, Bakewell, 1983.

[36] F.H. Harold, R.D. Child. Rangeland Ecology and
Management. Westview Press, Inc. New-York.
U.S.A, 1994.

[37] BCMF (British Columbia, Ministry of Forests).
2002. Considering tools for remediation.
Rangeland Health Brochure 4. British Columbia,
Canada, 2002, pp. 1-22

[38] L. Czeglédi, A. Radácsi. Overutilization of
Pastures by Livestock. GYEPGAZDÁLKODÁSI
KÖZLEMÉNYEK, 5 (2005) 29-35

[39] Marañón T. Biodiversidad de las comunidades
vegetales: escalas y components (Biodiversity of
plant communities: Scales and Components).
Proceeding of XXXVII Reunión Científica de la
SEEP, Sevilla-Huelva, Spain, 1997, pp. 15-24.

[40] R.W. Mc Dowell, J. J. Drewry, R. J. Paton.
Effects of deer grazing and fence- line pacing on
water and soil quality. Journal of Soil Use and
Management 20 (2004) 302-307.

[41] R. Evans. Overgrazing and soil erosion on hill
pastures with particular reference to the Peak
District. Journal of the British Grassland Society
32 (1977) 65-76.

[42] W.R. Teague, S.L. Dowhower, S.A. Baker, R.J.
Ansley, U. P. Kreuter. Soil and herbaceous plant
responses to summer patch burns under

continuous and rotational grazing, J. of
Agriculture, Ecosystems and Environment. 137
(2010) 113–123.

[43] H. Azarnivand, A. Farajollahi, E. Bandak, H.
Pouzesh. Assessment of the Effects of
Overgrazing on the Soil Physical Characteristic
and Vegetation Cover Changes in Rangelands of
Hosainabad in Kurdistan Province, Iran. Journal
of Rangeland Science, 1(2) (2011) 95-102

[44] M.A. Mohamed Saleem. Nutrient balance
patterns in African livestock systems,
Agriculture, Ecosystem and Environment 71
(1998) 241-254

****

4

Lunar Cycle based Cropping Calendar

Summary

The role of indigenous traditional knowledge in climate change adaptation and mitigation was highly and widely acknowledged and acclaimed. The objective of the current study was to document the traditional knowledge of Galo tribes of North Eastern Himalayan region of India regarding their cropping calendar based on lunar cycle. Scientific investigation was done to validate the belief. Phenomenological approach was used to study the knowledge system and sex pheromone traps are used to study insect population and activities. Validating the traditional belief, the catches of all the three groups of insects were found to be highest in the vicinity of Full Moon and lowest in the vicinity of New Moon. It was concluded that the validated and significant knowledge should be blended with modern policies and actions to enhance their wide acceptability and success.

INTRODUCTION

Indigenous peoples and their traditional knowledge play critical roles in the collective global responses to the challenges posed by climate change. The IPCC's Fourth Assessment Report (AR4) [1] noted that 'indigenous knowledge is an invaluable basis for developing adaptation and natural resource management strategies in response to environmental and other forms of change [2]. Also, UNFCCC (2011) [3] recognizes the conservation of traditional knowledge as co-benefits of ecosystem- based approaches to adaptation. The Institute of Advance Studies at the UN University recently identified more than 400 examples indicating the role of indigenous people in climate change monitoring, adaptation and mitigation[4]. Their traditional knowledge of ecosystem and their resilience to the vagaries of weather conditions are critical to building appropriate local to global responses. Despite of wide acknowledgement of traditional knowledge systems, their incorporation in the programmes and policies related to climate change mitigation and adaptation strategies are very limited [5]. How to apply this indispensable knowledge in participatory research to these new realities of climate change needs to be explored. The decision-makers must base their policies and actions on best available knowledge.

There are five types of traditional knowledge identified for climate change adaptation and mitigation- Resilient Properties, Plant Breeding, Wild Crop Varieties, Farming Practices and Climate Forecasting [6], and here we focus on Farming Practices. The underlying objective of this study was to understand local indigenous knowledge system, and to identify and strengthen their coping strategies for climate change and scale it up by blending it with modern scientific approach. The application of traditional knowledge in agricultural sector especially biological control of pests and diseases and ecological agriculture often enhances adaptability towards climate change [7]. In this study we present some significant decision making approaches of indigenous *Galo* tribes of Eastern Himalayan region regarding their cropping calendar based on lunar cycle. *Galo* tribes are indigenous community of West Siang District of Arunachal Pradesh located in North-Eastern Himalayan region of India. Their crop schedule depends on lunar cycle which influences most of their farming activities, based on nature of crops and inter-cultural practices needed. They believe that the cycle of farming activities must coincide with cycle of natural systems especially cosmic cycles for optimum production with minimum resource input. The primary concept of their belief is that the insect activities and dynamics alter with lunar cycle.

It is however important to note that not all indigenous practices are beneficial to the sustainable development of a local community; and not all indigenous knowledge can a p*riori* provide the right solution for a given problem [8]. Thus proper scientific validation and investigation must be employed before adopting any indigenous knowledge and integrating it into developmental programmes. To validate the indigenous belief, correlation study was done between the lunar cycle and insect dynamics.

MATERIAL AND METHODS USED

Study site

The indigenous knowledge was collected from farmers of *Gori, Soi* and *Bam* villages of West Siang District located in Eastern Himalayan region of Arunachal Pradesh. The research was done in ICAR Research Farm Gori, Basar, which is located at 27°58.590´ N latitude and 94°41.120´ E longitude at an altitude of 660 m msl.

Indigenous Knowledge System

To study the indigenous knowledge system, the phenomenological approach was used. The approach was chosen principally because it analyses natural behavior, as the indigenous communities perceived it rather than imposing any sort of external value judgment [9]. The Approach provides a crucial means for investigation in relation to how local people as "insider" come to know about some phenomenon [10]. The approach also helps to differentiate *noumena* (things as they are) from *phenomena* (things as we perceive them). So first hand information was collected through interactions with farmers various groups in their field. The strategy helped to gain understanding of the vitality of the indigenous methods.

The lunar cycle

The lunar cycle/phase was studied as proposed by to Nowinszky et al (2010) [11]. The phase angle values of the Moon have been downloaded from various internet sites. The 360° phase angle of the full lunar month (lunation) was divided into 30 divisions. The division in the ± 6° phase angle value vicinity of a Full Moon (0°, or 360°) was denoted as: 0, and starting from that, divisions in the direction of the First Quarter until the New Moon were denoted as -1, -2, -3, -4, -5, -6, -7, -8, -9, -10, -11, -12, -13, -14. Also starting from the Full Moon, divisions in the direction of the

Last Quarter until a New Moon were named: 1, 2, 3, 4, 5, 6, 7, 8, 9, 10, 11, 12, 13, and 14. The division including the New Moon was named: ±15. All divisions include 12 phase angle values.

Insect dynamics

Sex pheromone traps of most active insects, *halicoverpa armigera, spodeptra litura and fruit fly (Drosophila melanogaster)* were placed uniformly in 45 different locations (15 of each category) within the research farm (approximately area of 50 ha). Daily counting of trapped insects was done in early morning (6.00 AM). Finally the catch per day was assigned to the phase angle of the moon and average catch was plotted against each division.

Statistical Analysis

All statistical analysis and plotting were done using Microsoft Excel Software.

DESCRIPTIONS

Knowledge System

The date of sowing, inter-cultural operations (transplantation, manure applications, weeding etc) and harvesting of the *Galo* tribes are dictated by lunar cycle. These dates are different for different crops depending upon their duration and season. They believe that by choosing crop-specific date of sowing based on lunar cycle ensures, in most of the cases, that critical stages of crop does not coincide with period of high pest activity. They believe that the insect population and activity are highest in Full Moon and lowest in New moon. They also believe that the soil moisture content usually remains high during full moon and the chances of rain increase in vicinity of full moon. So sowing of seed must be done 4-5 days ahead of Full Moon day to avoid insect infestation and facilitate proper germination. Consequently, after 10-15 days when the plant reach critical growth stage, the growth is favored by low insect activity and population as the period coincide with days near to New Moon day.

Validation

In each case each insects depicts almost similar trends. The average insect catch was found to be highest during Full Moon and comparatively lowest in New Moon. There found to be positive correlation between the insect activities and lunar phase/cycle with correlation coefficient of r=0.87, r=0.82 and r=0.86 for *Halicoverpa, Spodeptra and Drosophila* respectively.

DISCUSSION

Moon continues to influence the daily activities of people around the world since ancient time. The impact of moon on origin and sustenance of life on earth was widely recognized and documented [12,13]. Studies suggested that the continuous cycles of wetting and evaporation along the shorelines of the early oceans due to tides caused by moon's gravitational attraction might have provided the kind of environment in which protonucleic acid fragments could begin to associate and assemble molecular strands leading to the origin of life [14]. In early Indian astronomy, moon was regarded as the mind of earth [15], controlling all the biological activities on earth.

Most of the farmers throughout the world take moon into account in their works, consciously or unconsciously, related to agriculture and allied activities [16]. In Ecuador it is the general custom not to till the soil, plant seed and harvest crops during New Moon [16].

The impact of lunar phase/cycle on the insect flight activity and dynamics was well documented [17,18,19]. It was found during the current validation process that the catch in the pheromone trap for each group of insect was invariably highest in the vicinity of Full Moon day and lowest in the vicinity of New Moon day. This finding was in agreement with finding of previous study [20], where during 15 years study in Texas, the flight activities of *Helicoverpa zea* and *Heliothis virescens* found to have significant positive correlation between the catch and the percentile value of lunar illumination. The highest catch was found during Full Moon (71%) and lowest during New Moon day (9%). Also the studies on Coconut Rhinoceros Beetle (*Oryctes rhinoceros L.*) found that the activity of the beetle is more during a Full Moon [21]. However, some of the previous studies contradict our result and observed no effective difference between the catch during Full Moon and New Moon [22,23]. This may be due to other location specific environmental factors dominating the dynamics of insects. Thus our scientific investigation satisfactorily supports the traditional belief system of the *Galo* tribes.

Adaptation to climate change includes all adjustments in behavior or economic structure that reduce the vulnerability of society to changes in the climate system [24]. Though climate change and global warming are global phenomenon, their impacts are very location specific. So location specific approaches are very important and need to be considered in all policies and recommendations. The policies and actions without the consensus of local indigenous people often led to poor participation and fail to achieve and realize its full potential [25,26,27]. Integration of indigenous knowledge into climate change mitigation and adaptation policies can lead to the development of effective and robust strategies in a very cost-effective, participatory and sustainable way [28,29]. To concludes, the traditional knowledge system must complement rather than to compete with modern scientific knowledge.

Incorporation of these jewels of knowledge, preceded by proper validation, could significantly strengthen policies and recommendations for sustainable development especially under the scenario of climate change. Proper identification and documentation of traditional knowledge must be done to acknowledge and give due recognition to the traditional knowledge holder [30], beside its circulation for global co-benefits.

REFERENCES

1. IPCC, *Summary for Policymakers, Fourth Assessment Report (AR4)*. New York, Cambridge University Press (2007)

2. Parry M.L., Canziani O.F., Palutikof J.P., van der Linden P.J. and Hanson C.E., (eds.) *Contribution of Working Group II to the Fourth Assessment Report of the Intergovernmental Panel on Climate Change*. Cambridge, UK and New York, Cambridge University Press (2007)

3. UNFCCC, Ecosystem-based approaches to adaptation: compilation of information. Note by the Secretariat. Available at < http://unfccc.int/resource/docs/2011/sbsta/eng/inf 08.pdf> (2011)

4. Galloway McLean, K., 2010. Advance Guard: Climate change impacts, adaptation, mitigation and indigenous peoples. A compendium of case studies. UNU-IAS (2010)

5. Below T., Artner, A., Siebert, R. and Sieber, S., Micro-level Practices to Adapt to Climate Change for African Small-scale Farmers. A Review of Selected Literature. IFPRI Discussion Paper 00953. http://www.ifpri.org/site/default/flies/publications /ifpridp00953.pdf (2010)

6. Swiderska K., Reid H., Song Y., Li J., Mutta D., Ongogu P., Mohamed P., Oros R., and Barriga S., *The role of traditional knowledge and crop varieties in adaptation to climate change and food security in SW China, Bolivian Andes and coastal Kenya.* Paper prepared for UNU-IAS workshop Indigenous

115

Peoples, Marginalised Populations and Climate Change: Vulnerability, Adaptation and Traditional Knowledge, Mexico, July (2011)

7. FAO, FAO and Traditional Knowledge: The Linkages with Sustainability, Food Security and Climate Change Impacts. Economic and Social Development Department. Viale delle Terme di Caracalla - 00153 Rome, Italy (2009)

8. Ajani E.N., Mgbenka R.N. and Okeke M.N., Use of Indigenous Knowledge as a Strategy for Climate Change Adaptation among Farmers in sub-Saharan Africa: Implication for Policy. *Asian Journal of Agricultural Extension, Economics & Sociology*, 2(1), 23-40 (2013)

9. Mugati T. and Maposa R.S., Indigenous weather forecasting: A phenomenological study engaging the shone of Zimbabwe. *The Journal of Pan African Studies*, 4(9), 102-111(2012)

10. Cox J.L., An Introduction to the Phenomenology of Religion, Gweru: Mambo Press (1992)

11. Nowinszky L., Barczikay G. and Pukas J., The Relationship between Lunar Phases and the Number of Pest Microlepidopter Specimens Caught by Pheromone Traps. *Asian Journal of Experimental Biological Sciences*, 1(1), 14-19 (2010)

12. Munro Fox H., Selene or Sex and the Moon. Kegan Paul, Tench, Trubner & Co. London (1928)

13. Spradley J.L., Ten lunar legacies: Importance of the Moon for the on Earth. *Perspectives on Science and Christian Faith*, 62(4), 267-275 (2010)

14. Lathe R., "Fast Tidal Cycling and the Origin of Life," Icarus 168, no. 1, 18–22 (2003).

15. Krishnamacharya E., Spiritual Astrology. Kartik Printers, Madras, India (1983)

16. Carlier H. The moon and agriculture. *Ileia Newsletter*, 3(1), 22 (1987)

17. Vaishampayan, S.M., New design of light trap for survey and management of insect pest population in agro and forestry ecosystems. *Indian Journal of Extension*, 44, 201-205 (1982)

18. Nag A. and Nath P., Effect of moon light and lunar cycle on the light trap catches of cutworm, Agrotis ipsilon (Hufn) moths. *Journal of Applied Entomology*, 111, 358-360 (1991)

19. Mishra P.N., Singh M.P. and Nautiyal M.C., Effect of Moon Light and Lunar Periodicity on the Attraction of Black Cutworm Moth *Agrotis flammatara* (Schiffer-Mueller) on Light Trap. *Pertanika Journal of Tropical Agricultural Science*, 22(1), 69-72 (1999)

20. Parajulee M. N., Slosser J. E. and Boring E. P., Seasonal activity of *Helicoverpa zea* and *Heliothis virescens* (Lepidoptera: Noctuidae) detected by pheromone traps in the rolling plains of Texas. *Environ. Entomol.*, 27 (5), 1203- 1219 (1998)

21. Kamarudin N. and Wahid M.B., Immigration and activity of *Oryctes rhinoceros* within a small oil palm replanting area. *J. Oil Palm Res.*, 16 (2), 64-77 (2004)

22. Suckling D. M. and Brown B., Daily performance of orchard pheromone traps. Proceedings of the Forty

Fifth New Zealand Plant Protection Conference, Wellington, New Zealand. pp. 279-284 (1992)

23. Sekhar P. R., Venkataiah M. and Venugopal, Rao, N., Effect of moon light and lunar periodicity on the pheromone trap catches of *Helicoverpa armigera* (Hubner) moths. *Annl. Agri. Res.*, 17 (1), 53-55 (1995)

24. Smith J.B., Ragland S.E. and Pitts G.J., A process for evaluating anticipatory adaptation measures for climate change. *Water, Air and Soil Pollution*, 92, 229–238 (1996)

25. Howes M., The use of indigenous technical knowledge in development. In: Brokensha, D.W., Werner, O. and Warren, D.M. (Eds.). Indigenous knowledge systems and development. University Press of America Inc., Lanham, MD (1980)

26. Woodley E., Indigenous ecological knowledge systems and development. *Agric Human Values*, 8, 173-178 (1991)

27. Nyong A.O. and Kanaroglou P.S., Domestic water demand in rural semi-arid North-eastern Nigeria:

119

Identification of determinants and implications for policy. *Environ Plan.* 34(4), 145-158 (1999)

28. Hunn E., What is traditional ecological knowledge? In: Williams N. and Baines G. (Eds.). Traditional ecological knowledge: Wisdom for sustainable development. Centre for Resource and Environmental Studies, ANU, Canberra. 3-15(1993)

29. Robinson J. and Herbert D., Integrating climate change and sustainable development. *International Journal of Global Environmental Issues*, 1(2), 130-148 (2001)

30. Tripathi S.K., Traditional Knowledge: Its significance and Implications. Indian Journal of Traditional Knowledge, 2(2), 99-106 (2003)

****

5

Seasonal Crop Calendar and Gender Disaggregated Daily Activities

Summary

Seasonal calendar and daily farming activities are vital in understanding the farming system of a community and identifying the period of intensive agriculture and lean period for better resources management and planning. Also the knowledge of gender roles within a community can provide practical guidance for policy makers and researchers involved in technology investment, extension services and marketing interventions. The objective of the current study was to evaluate seasonal calendar and gender disaggregated daily activities of indigenous Galo farmers of Arunachal Pradesh. Phenomenological approach was used to understand the system and time use survey was done to document the time spent by men and women in different activities. The result reveals rice-based farming system of the tribe with seasonal calendar of major crops that is guided by the climate of the location. Also, it was found that farming is a family activity of the community with well defined task for each family member based on gender. Women plays major role in the farming and also have vital responsibility in preservation and conservation of indigenous local crop varieties and seeds for next season. Generally it was seen that the labor burden of Galo women exceed than that of their male counterpart. Beside agricultural field activities, women are exclusively responsible for livestock care, marketing, household activities including child care and firewood collection. Keeping in view the crucial role of women in agricultural activities, gender informed approaches in agricultural policies, discussions, strategies and programs are needed for holistic development of the sector.

INTRODUCTION

Arunachal Pradesh is situated in the Eastern Himalayan region of India. It is the home of 26 major tribes and 110 sub-tribes that are indigenous to the state [1]. *Galo* tribe is one of the major tribes in the East Central part of the state.

Agriculture is the mainstay of population of Arunachal Pradesh with above 80 per cent of the total population directly or indirectly depends on it for their livelihood [2]. Predominantly shifting cultivation (*jhum*) is practiced in more than 50 per cent of grossed cropped area. Agriculture sector in the state is underperforming and not able to achieve its full potential due to difficult agro-climatic condition with undulating topography, poor soil quality, subsistence nature of agriculture, inadequate investment capabilities and improper enterprise mix. These factors are compounded by change in precipitation patterns and temperature due to climate change. To address the issue concentrated research initiatives and policies are needed tailored to the local need and preferences, and must be based on their farming system and socio-economic life.

Each tribe of the state has their own seasonal calendar based on the climate of the location. As seasonal calendar indicates month wise activities with regards to agriculture and animal husbandry, it is very crucial in understanding the farming system of a location or community to customize policies and research. The seasonal calendar may also help to identify period of intensive agricultural and lean period for better management of resources. Also for majority of the tribes, farming is a family or group activity where each member has certain defined tasks mainly based on their respective genders. Understanding the gender division of labor by crop and tusk is crucial on many levels while shaping the structure of development assistances and in deciding the target group. Studies have shown that failure to recognize the different roles of men and women is a big hurdle in development because it results in misguided projects and policies leading to poor productivity and nutritional insecurity [3]. The objective of the current study was to understand the crop calendar of *Galo* tribes and their gender disaggregated daily activities. This was an initiative to understand the agricultural system of the location. The gender disaggregated daily activities reflects the level of involvement of each gender in farming and the socio-economic life of the community and time spent by men and women in different activities. However, the role of women is particularly significant as they are more focused towards livelihood improvement of their family and the society, and sustainability of the ecosystem. Also, they are more concerned with development of agricultural practices which were affordable, use local inputs, integrated to other facets of life, practical and result oriented [4]. In the current study role of women is mainly evaluated because in tribal communities the invisibility of contribution of tribal women farmer in socio-economic life is a very challenging issue.

MATERIALS AND METHODS

Study site

The study is mainly concentrated in the Basar and nearby villages in the West Siang district of Arunachal Pradesh around the geographical coordinates of N 27°59.27′ and E 94°41.26′ in the Eastern Himalayan region of India.

Method

To study the agricultural system of the *Galo* tribes, the phenomenological approach was used. The approach evaluates and analyses natural behavior and instinct, as the indigenous communities perceived it rather than imposing any sort of external value judgment [5]. The approach also helps to differentiate *noumena* (things as they are) from phenomena (things as we perceive them). So the first hand information was collected through interactions with farmers of different age groups and gender. The metaphors, folklore and proverbs that give better prospective were collected and studied to get better knowledge and inside on the subject. Key informants from different villages were selected including village elders, youths, women and priests. Time use survey [6] was done to study the daily activities including time spent in farming, household activities, social behavior etc. of both men and women in the community.

THE SYSTEM

The Galo tribes have very well defined seasonal crop calendar distributed throughout year. Their seasonal calendar is guided by the climate of the location. They generally have rice-based farming system with major crop includes *jhum*/upland rice, lowland rice, maize, mandarin oranges, banana, tuber crops and ginger & turmeric.

They involve in *jhum* cultivation throughout the year which begins in the month of December-January with jungle clearing followed branch cutting in February, cutting and stubble collection and burning in March, dibbling in April and May, weeding during June-July-August, panicle harvesting and drying during September and October, threshing and winnowing during November-December. All *Jhum* related activities are mainly guided by different natural indicators [7]. The wetland rice cultivation generally starts in the month of April with clearing of field, nursery preparation in the month of May, land preparation and transplanting in the months of June-July, weeding during August-September-October, panicle harvesting generally during November and threshing & winnowing in December. Maize is another important crop of the community and is sown twice a year during March-April and August-September with sowing of one crop is preceded by harvesting of the previous crop. The main horticultural crop is mandarin orange and new orchard is generally planted in month of March and transplanted during June with onset of rainfall. Harvesting of orange generally starts in the month of November and December. Land preparation and planting of banana is generally done in the month of June with onset of monsoon and harvested in the month of August. Tuber crops like colocasia, elephant foot yam, tapioca etc are important component of food of the indigenous *Galo* tribes. Generally the sowing of tuber crops are done in the months of March-April with pre-monsoon shower and harvested in the months of October-November. Ginger and turmeric are main commercial crops of the location. Generally sowing of ginger and turmeric starts in the months of April-May and harvesting in the months of January-February. Pig and mithun (*Bos frontalis*) are major among livestock and reared throughout the year. The rearing of livestock is constrained by seasonal diseases. Generally the swine fever in pig appears in the months of June-July and November-December. The foot and mouth disease of the cattle generally were seen in the months of January-February and June-July. The seasonal calendar occasionally deviates from its normal dates due to variation in the onset of rainfall and due to change in weather conditions. The indigenous communities change their seasonal activities as per the perceived weather conditions which they predict using different traditional methods by observing behavior of animals, insects, birds etc and plan their sowing and harvesting schedule as per lunar cycle.

The daily activities of a *Galo* rural family is governed by the agricultural activities as shown in Table 2. For typical Galo women, the working hour is normally 17 hours from 3:30 AM to 8:30 PM. As per the daily activity (Table 2), women spent 51 per cent of total working hours in agricultural activities including field works, livestock care, selling the surplus in local market etc and 41.2 per cent in household activities including cooking, cleaning, firewood collection, etc. Beside these, they also need to take care of the elderly & sick members and children of the family. They are also responsible for keeping the harvested grains and seed for next crop. While for men the working hour is normally 16 hours. Men normally spent 50 per cent of time in agricultural field during cropping season. The activities that require hard labor like ploughing, jungle cutting, field preparation and mithun rearing have been carried out by the men. On an average they spent 6 per cent of time in household activities and 6 per cent in social activities. However in some occasion they need to spend more time in social activities like village court.

The Crop Calendar is a tool that contains information on planting, sowing and harvesting periods of locally adapted crops in specific agro-ecological zones [8]. Seasonal Crop Calendar supports farmers in taking appropriate decisions on crops and their sowing period keeping in view the agro-ecological dimension and climate. It helps the researchers and policy makers to understand the farming system of the location and device need-based techniques and measures to improve the agriculture and livelihood of the targeted community. It also provides solid base for preparation of contingency crop planning and emergency planning of the rehabilitation of farming systems after disasters. In Australia seasonal calendars of the aboriginal *Ngadju* helps identify changes in the climate and in combating challenges [9]. Seasonal calendar were found to represent a wealth of indigenous ecological knowledge.

Regarding participation of men and women in farming, generally the role of women is found to be more significant. The activities like sowing, weeding, harvesting and grain separation has been done by the women. The women folks are also involved in forest cutting and firewood collection throughout the year in addition to other regular household activities. Mostly the women take care of the livestock and poultry. The women are also involve in selling of the surplus harvest to the local market, and engage in trade and marketing. They are responsible for storing seeds for next harvest. The women also collect wild edible foods and medicinal herbs from the nearby forest.

The Adi-Galo women are excellent gatherer of wild food and herbs and have unmatched skill of indentifying the edible plants, tubers, edible mushrooms and herbs. They generally go in groups to the forest to collect sufficient supplies for the whole family. Table 1 shows some most commonly used wild plants as vegetables, condiments and spices by the Adi-Galos. They provide important ingredients of their daily diet and the activity is a vital part of the Adi-Galo economy.

Table-1. Some most commonly used wild vegetables, condiments and spices by the Adi-Galos.

Botanical Name	Local Name
Allinia galangal (Linn.) Wild. (Zingiberaceae)	*jakur*
Clerodendron Colebrookianum Linn. (Verbenaceae)	*Oin-taap*
Dioscorea esculenta (Lour.) Burkill (Dioscoreacea)	*Iqin-tabu*
Diplazium Esculentum Retz	*o-takaa*
Fagophyllum esculentum Moench (Polygonaceae)	*okuuu*
Ficus caudate Wall (Moraceae)	*takuk*

Fragaria indica (Linn.) Wall. *(Rosaceae)*	*enci*
Ipomoea batata (Linn.)Lam. *(Convolvulaceae)*	*maalww*
Manihat Esculenta	*hwwn-iqin*
Pouzolzia Bennettiana	*oik*
Solanum Nigrum Linn.	*orv*
Spilanthes Oleracea (Linn.) *D.C(Asteraceae)*	*marsaa*
Zanthoxylum acanthopodium *(Linn.)DC(Rutaceae)*	*oxor*

(Source: Tarak, Koyu, Samal and Singh 2009)

The women's role in agriculture and allied activities were found to be more prominent in *Galo* tribes. The labor burden of tribal women generally exceeds than that of their male counterpart as it also includes household responsibilities including preparation of food, child care, collection of firewood etc. Similar roles of women in agricultural activities are also observed in the tribal society of Gujarat [10]. Studies show that the role of women in agricultural activities is around 60-80 per cent in Gambia, over 32 per cent in India as a whole and over 50 per cent in China [11]. Several studies based on time-allocation have found that women work significantly more than men when household activities, child rearing etc were included [12,13,14]. Women in the indigenous tribal community were found to play significant role in the conservation of germplasm of various indigenous crop varieties [15,16]

. Studies indicated that rural women usually oversee small household livestock and sometime cattle throughout the world [17]

. Women found to have better knowledge of the medicinal plant and different tree species [18]

. Also, in Nepal it was found that role of women is especially prominent in hill regions where stronger crop-animal-tree integration requires greater involvement [19]

. The role of women towards agriculture has never been static and it changes with the changing need, environment and context·

CONCLUSION

Agriculture is central to the livelihood of the rural resource poor community and it can be engine to socio-economic development. So the developmental projects and initiatives must be well focused to the customized need to the targeted community. Seasonal crop calendar is benchmark for any research and policy design.

Change in the seasonal calendar indicates the change in the agro-climate of the location. In future work such changes could be studied. The gender disaggregated daily activities of a community helps us to understand the socio-economic structure of the community. Women in the tribal community clearly have central role to play in achieving potential agricultural productivity and development. Despite of crucial role of women in agricultural activities, gender based approach in agricultural policies, discussions, strategies and programs are largely missing. The women should be equipped with technological skill and know-how. The changes, both in climate and economic structure, may increase vulnerability of individuals with few resources, especially rural women with limited access to crucial services and opportunities. Thus, efforts are need to be made to enable women to move beyond current production for subsistence to production under commercial mode. Secondary agricultural practices like beekeeping, mushroom cultivation, backyard poultry etc should be promoted among the indigenous community to make income during the event of failure of major crop due to weather related extreme events.

REFERENCES

1. Srivastava, R.C. and *Adi* Community. Structure of Jhum (Traditional Shifting Cultivation System): Prospect or Threat to Climate. *Indian Journal of Traditional Knowledge*, **8**(2): 146-153 (2009).

2. SAPCC. Arunachal Pradesh State action plan on climate change. Department of Environment & Forest, Govt. Of Arunachal Pradesh, Itanagar (2011). www.nicra-icar.in

3. Mehra, R. and Rojas, M.H. Women, Food Security and Agriculture in a Global Marketplace. A

Significant Shift. International Centre for Research on Women. Washington (2008).

4. Lopez, Victoria M. and Custro, M. Women's Wisdom: Documentation of Women's knowledge in Agriculture (case studies from Philippines, Thailand and Pakisthan). Pesticide Action Network Asia and Pacific. Penang, Malaysia (2012).

5. Mugati, T. and Maposa, R.S. Indigenous weather forecasting: A phenomenological study engaging the shone of Zimbabwe. *The Journal of Pan African Studies*, *4*(9): 102-111 (2012).

6. Ironmonger, Duncon (1999), " An Overview of Time Use Surveys", Households Research Unit, Department of Economics, The University of Melbourne, Australia

7. Gupta, V. *Jhum* cultivation practices of the Bangnis (*Nishis*) of Arunachal Pradesh. *Indian Journal of Traditional Knowledge*. 4(1): 47-56, (2005).

8. FAO. 2010. http://www.fao.org/agriculture/seed/cropcalendar/welcome.do

9. O'Connor, M.H. and Prober, S.M. A calendar of *Ngadju* seasonal knowledge. A report to Ngadju Community and Working Group. CSIRO Sustainable Ecosystems, Floreat, WA. (2010).

10. Prajapati, R.R., Thakkar, K.A. and Prajapati, M.R. Participation of Tribal Farm Women in Agriculture Related Indigenous Resource Management Activities

in Banaskantha District of Gujarat. *Journal of Community Mobilization and Sustainable DeveloPMent.* 7(1): 21-25 (2012).

11. Terri Raney, Gustavo Anríquez, Andre Croppenstedt, Stefano Gerosa, Sarah Lowder, Ira Matuscke, Jakob Skoet and Cheryl Doss. The role of women in agriculture. ESA Working Papers No. 11-02. Agricultura Development Economics Division. FAO. http://www.fao.org/publications/sofa/en/ (2011).

12. Ilahi, N. 2000. The intrahousehold allocation of time and tasks: What have we learnt from the empirical literature? Policy Research Report on Gender and Development, Working Paper Series No. 13, World Bank: Washington DC.

13. Kes, A. and H. Swaminathan. 2006. Gender and Time Poverty in Sub-Saharan Africa. Chapter 2 in Blackden, C.M. and Q. Wodon. (Eds.). Gender, Time Use, and Poverty in Sub-Saharan Africa. World Bank Working Paper No. 73, The World Bank, Washington, D.C.

14. Budlender, D. 2008. The statistical evidence on care and non-care work across six countries. Geneva: United Nations Research Institute for Social Development (UNRISD).

15. Vartak, K.V.D. Sacred groved of tribals for in-situ conservation of biodiversity. In S.K.Jain (eds.) Ethnobiology in Human Welfare, pp. 300-302 (1996).

144

16. Rangalakshmi et al. Rural and Tribal Women in Agro-biodiversity Conservation: Indian Case Study. RAP Publication 2002/08, MSSRF, Chennai. FAO Regional Office for Asia Pacific, Bangkok, Thailand (2002)

17. Bravo-Baumann, H. Capitalization of Experiences on the contribution of Livestock Projects to Gender Issues. Working Document, Swiss Agency for Development and Cooperation, Bern (2000). http://www.fao.org /WAIRDOCS/LEAD/X6106E/x6106e01.htm#TopOf Page.

18. Shiva, V. and Dankelman, I. Women and Biological Diversity. Growing Diversity, Genetic Resource and Local Food Security, A.O. David Cooper (eds), London: *Intermediate Technology Publication* (1992).

19. Timsina, D., Timsina, J. and Chhetry, M.B. Womens' role in Nepalese farming systems: a comparative study of a hill and an inner tarai farming system site, Working Papers on Women in International Development No. 193, Michigan State University, Office of Women in International Development (1989).

6

Indigenous Traditional Knowledge of Weather Prediction using Songs of Annual Cicadas

Summary

Weather forecasting based on the observation of the natural phenomenon always been subject of interest for the indigenous farming communities from time immemorial. The objective of the current study was to identify and validate weather prediction technique of indigenous Galo tribes of Arunachal Pradesh using songs and singing behavior of annual cicada. Phenomenological approach was used to study the belief; and for scientific investigation and validation Bright Sunshine Cards were compared with the noise graph of the cicada. Comparison reveals positive correlation between the singing behavior of the cicada and forecast as per the belief of the indigenous community. Continuous loud singing indicates clear/dry weather for next 2 to 3 days, while sudden silence indicate that the rain in next 1 or 2 days.

INTRODUCTION

The knowledge of probable short term weather thus, always has been vital to stabilize yield through management of resources and inputs. The information helps farmers and the planners to exploit the potential of good weather and minimize the impact of bad weather. Besides, it also helps farmers to avoid the adverse effect of weather events like heavy rainfall, dry spell, high wind speed which may influence the crop productivity. Weather prediction is becoming an essential component of agricultural activity especially in the current era of climatic variability due to climate change.

Weather always been the subject of study and observation for the indigenous communities from the antiquity, especially for those related to agriculture and allied sectors. There are huge repository of indigenous knowledge related to weather forecasting that are reflected from their culture, festivals, rituals, folklores and beliefs [1,2]. For instance, the farmers and fishermen in the Philippines rely on change in the behavior of animals and insects to predict the possibility of rains and bad weather. Indigenous knowledge, being the outcome of observation of several generations, has close relationship with the environment and the ecosystem. Indigenous Traditional Knowledge in broad sense is knowledge developed and holds by the original inhabitants of an area and their use of it in daily life [3]. By understanding to what people observe in their natural environment and how these observations are grouped and interpreted, we gain insight into the ways in which people think about and interact with their local environment. As traditional knowledge is a cultural tradition that is constantly being developed and adjusted and transmitted from generation to generation, they are widely accepted among the community and most of them have some scientific implication behind them.

Currently we are witnessing a reversal of the trend and a return towards traditional methods that rely on local technical know-how and practices, for climate resilient and location–specific sustainable agriculture to combat the challenges posed by climate irregularities due to global warming and soil degradation. The indigenous knowledge should be ensured, encouraged and facilitated to develop coordination between modern scientific techniques and local environment. The IPCC's Fourth Assessment Report (AR4) [4] noted that 'indigenous knowledge is an invaluable basis for developing adaptation and natural resource management strategies in response to environmental and other forms of change [5].

Also, UNFCCC (2011) [6] recognizes the conservation of traditional knowledge as co-benefits of ecosystem- based approaches to adaptation. The indigenous knowledge system were in fact accumulated incrementally, tested by trial-and-error and transmitted to future generations orally or by shared practical experiences, thus are sustainable and socially acceptable [7]. The need of the hour is to understand and validate such knowledge and incorporate them with the conventional scientific techniques. Coupling of these techniques with modern strategies would enhance mass enthusiasms in mitigation and adaptation strategies.

The objective of the current study was to identify one of the most effective indigenous techniques of weather forecasting for next 2-5 days by using the songs of annual cicada practice by the *Galo* tribes of West Siang district of Arunachal Pradesh in the Eastern Himalayan region of India. Farming is the main occupation of almost ninety percent population of the area The people live very close to nature and are very keen observer of nature and its phenomenon. The indigenous technique was validated by comparing with the actual weather data recorded in the observatory.

STUDY SITE

The district West Siang of Arunachal Pradesh is located between 93°57'- 95°23' E latitude and 27°69'- 29 ° 27' E longitudes with an altitude range between 150 to 6000 metres above mean sea level. The district of West Siang comes under the Eastern Himalayas, warm per humid eco-region under zone two.

METHODS USED FOR THE STUDY

The phenomenological approach was used to study the various cultural system associated with the technique. The approach was chosen principally because it evaluates and analyses natural behavior and instinct, as the indigenous communities perceived it rather than imposing any sort of external value judgment [1]. The means and techniques through which the local people as "insider" come to know about some phenomenon can be investigated through this approach. The approach also helps to differentiate *noumena* (things as they are) from phenomena (things as we perceive them). So the first hand information was collected through interactions with farmers of different age groups. The metaphors, folklore and proverbs that give better prospective on the traditional knowledge were collected by us and studied.

SCIENTIFIC INVESTIGATION

The noises of the Cicadas were recorded in their natural habitat with ordinary sound recorder system, mostly mobile phones so that it can be easily downloaded in the computer. Each day from different locations the noise were recorded simultaneously and stored in the computer date wise. The loudness of the noise was studied simply with Windows Media Player and the snapshots of the graphs were recorded and placed date wise. The bright sunshine hours of each day was recorded with Cambell Stroke Sunshine Recorder at the Agro-meteorological Observatory. The

Sunshine Recorder Cards and the Noise Graphs were compared to identify any pattern or correlation. The weather forecast based on the indigenous belief was compared with the actual weather events.

Annual Cicada: Cicadas are flying, plant-sucking insects of the Order Hemiptera; their closest relatives are leafhoppers, treehoppers, and fulgoroids. Adult cicadas tend to be large (most are 25-50 mm), with prominent wide-set eyes and ocilli, short antennae, and clear wings held roof-like over the abdomen. The adult male cicada possesses two ribbed membranes called tymbals, one on each side of its first abdominal segment. By contracting the tymbal muscle, the cicada buckles the membrane inward, producing a loud click. As the membrane snaps back, it clicks again. The two tymbals click alternately. Air sacs in the hollow abdominal cavity amplify the clicking sounds. The vibration travels through the body to the tympani, which amplify the sound further. This produces a conspicuous acoustic signals or "songs" for which the cicadas are probably best known. Cicadas are the only insects capable of producing such a unique and loud sound. Some larger species can produce a call in excess of 120 decibels at close range. This noise or "song" is the basis of weather forecasting technique used by the traditional people of the area. They generally appear during first week of April and stay up to last week of October in the district.

FINDINGS

Annual cicada locally known as *"Goi"* is a very important insect for the farming communities among the galo tribes in the region as the unique song of the insect help them to get idea about the weather for next 2-5 days. The local people believe, with their generations of observation, that the noise duration and loudness of the insect is very much related to the approaching weather conditions. Their general beliefs are as under:

- Continuous loud noise for above 2 hours

  Prediction: Clear and dry weather for next 2-5 days with no possibility of rainfall.

- Continuous noise for days and nights for 2-3 days

  Prediction: No rain for a considerable period up to 15 days or more.

- When the cicada suddenly stop singing for the day:

  Prediction: There is possibility of rainfall in next 1-2 days

Scientific Investigation

The sample method employed to correlate the noise graph and the Sunshine hour. The noise graph signifies the loudness and frequency of the noise of the annual cicada recorded. To validate the forecast, the Bright Sunshine (BSS) hours recording cards for next 2-5 days were compared with the noise graph. The BSS card shows the bright sunshine hours indicating the atmospheric state. Taking the flexibility of prediction from 2-5 days, on visual comparison it has been found that in above ninety percent of the case there is positive correlation between the noise graphs and BSS cards. More the disturbance in the graph or more the loudness more is the burnt in the BSS cards indicating more hours of bright sunshine. While when there is silence for a day or two, in next 2-3 days either there is rainfall or no burnt in the BSS cards. From the total observations, we take 500 observations at a random for statistical analysis to verify the significance of the forecasting. It was found that out of 500 observations, 416 cases are in accordance with the expectation.

The scientific investigation reveals that the weather forecasting technique is not a mere by chance or a belief system. There might be some adaptations allowing cicadas to detect imminent changes in weather conditions. The change in the singing behavior might be some modified sexual behavior as found in most of the Hemiptera [8],

2013) in response to the change in the atmospheric conditions.

CONCLUSION

Indigenous knowledge, being the outcome of observation of several generations, is time tested for the area and are mainly eco-friendly and sustainable. So it is the need of the hour to understand and validate such jewels of knowledge and incorporate them with the conventional scientific techniques. As the cited technique proved very much helpful to predict the erratic weather of the area and take short time crop management decisions.

REFERENCES

1.	Mugati, T. and Maposa, R.S. Indigenous weather forecasting: A phenomenological study engaging the shone of Zimbabwe. *The Journal of Pan African Studies,* **4**(9): 102-111 (2012).

2.   Mapara, J. 2009. Indigenous Knowledge Systems in Zimbabwe: Juxtaposing Postcolonial Theory. *Journal of Pan African Studies*, vol.3, no.1, September 2009

3.   Prober, S. M., M. H. O'Connor, and F. J. Walsh. 2011. Australian Aboriginal peoples' seasonal knowledge: a potential basis for shared understanding in environmental management. *Ecology and Society* **16**(2):   12.   [online]   URL: http://www.ecologyandsociety.org/vol16/iss2/art12/

4.   IPCC, Summary for Policymakers, Fourth Assessment Report (AR4). (2007). New York, Cambridge University Press

5.   Parry, M.L., Canziani, O.F., Palutikof, J.P., van der Linden, P.J. & Hanson C.E. (eds.) (2007). Contribution of Working Group II to the Fourth Assessment Report of the Intergovernmental Panel on Climate Change. Cambridge, UK and New York, Cambridge University Press.

6.   UNFCCC, Ecosystem-based approaches to adaptation: compilation of information. Note by the Secretariat. Available   at   < http://unfccc.int/resource/docs/2011/sbsta/eng/info 8.pdf> (2011)

7.   Ohmagari, K. & Berkes, F. (1997). Transmission of indigenous knowledge and bush skills among the Western James Bay Cree women of subarctic Canada. Human Ecology 25: 197- 222

8.   Ana Cristina Pellegrino, Maria Fernanda Gomes Villalba Peñaflor, Cristiane Nardi, Wayne Bezner-Kerr, Christopher G. Guglielmo, José Maurício Simões Bento. Weather Forecasting by Insects: Modified Sexual Behaviour in Response to Atmospheric Pressure Changes. PLOS one. http://dx.doi.org/10.1371/journal.pone.0075004 (2013)

Author's Profile

Kaushik Bhagawati
Engaged in the National Innovation on Climate Resilient Agriculture programme of Indian Council of Agricultural Research and actively doing research on various aspects of climate and climate change especially related to approaches of Indigenous Communities. He is also involved in providing need-based advisories to the farmers of the state of Arunachal Pradesh through mobile sms services. He is author of several research papers published national and international journals, books and book chapters.
Email: kaushik.iasri@gmail.com

Kshitiz Kumar Shukla
Working as Senior Research Fellow under National Innovation on Climate Resilient Agriculture programme of Indian Council of Agricultural Research and actively involved in research activities related to impact of climate change on agriculture in molecular level. He is also closely associated with the study of behaviour of indigenous communities particularly related to *jhum* cultivation. He is author of several research papers published national and international journals, books and book chapters.
Email: kshukla65@gmail.com

Amit Sen

He is currently working as Assistant Professor in Mewar University, Rajesthan. He was actively involved in the climate change related activities during his previous service as SRF under NICRA project.

Email: amitsenbhl28@gmail.com

Rupankar Bhagawati

He is presently the Joint Director of ICAR Research Complex for NEH Region, Arunachal Pradesh Centre, Basar. He is a eminent scientist of plant pathology and closely associated with climate change studies.

Email: rbhagawati@gmail.com

www.ingramcontent.com/pod-product-compliance
Lightning Source LLC
Chambersburg PA
CBHW021428170526
45164CB00001B/146